SpringerBriefs in Molecular Science

Chemistry of Foods

Series editor

Salvatore Parisi, Industrial Consultant, Palermo, Italy

The series Springer Briefs in Molecular Science: Chemistry of Foods presents compact topical volumes in the area of food chemistry. The series has a clear focus on the chemistry and chemical aspects of foods, topics such as the physics or biology of foods are not part of its scope. The Briefs volumes in the series aim at presenting chemical background information or an introduction and clear-cut overview on the chemistry related to specific topics in this area. Typical topics thus include: - Compound classes in foods – their chemistry and properties with respect to the foods (e.g. sugars, proteins, fats, minerals, ...) - Contaminants and additives in foods – their chemistry and chemical transformations - Chemical analysis and monitoring of foods - Chemical transformations in foods, evolution and alterations of chemicals in foods, interactions between food and its packaging materials, chemical aspects of the food production processes - Chemistry and the food industry – from safety protocols to modern food production The treated subjects will particularly appeal to professionals and researchers concerned with food chemistry. Many volume topics address professionals and current problems in the food industry, but will also be interesting for readers generally concerned with the chemistry of foods. With the unique format and character of Springer Briefs (50 to 125 pages), the volumes are compact and easily digestible. Briefs allow authors to present their ideas and readers to absorb them with minimal time investment. Briefs will be published as part of Springer's eBook collection, with millions of users worldwide. In addition, Briefs will be available for individual print and electronic purchase. Briefs are characterized by fast, global electronic dissemination, standard publishing contracts, easy-to-use manuscript preparation and formatting guidelines, and expedited production schedules. Both solicited and unsolicited manuscripts focusing on food chemistry are considered for publication in this series.

More information about this series at http://www.springer.com/series/11853

Sara M. Ameen · Giorgia Caruso

Lactic Acid in the Food Industry

 Springer

Sara M. Ameen
Medical Research Laboratories, Faculty
  of Science
Helwan University
Cairo
Egypt

Giorgia Caruso
Industrial Consultant
Istituto Zooprofilattico Sperimentale
Palermo
Italy

ISSN 2191-5407   ISSN 2191-5415   (electronic)
SpringerBriefs in Molecular Science
ISSN 2199-689X   ISSN 2199-7209   (electronic)
Chemistry of Foods
ISBN 978-3-319-58144-6   ISBN 978-3-319-58146-0   (eBook)
DOI 10.1007/978-3-319-58146-0

Library of Congress Control Number: 2017939892

Printed on acid-free paper

This Springer imprint is published by Springer Nature
The registered company is Springer International Publishing AG
The registered company address is: Gewerbestrasse 11, 6330 Cham, Switzerland

Contents

Chapter 1
The Importance of Lactic Acid in the Current Food Industry. An Introduction

Abstract The aim of this chapter is to give a brief and reliable overview of possible uses for lactic acid as additive in the current food industry. The importance of lactic acid in the food industry is apparently linked to microbial activity of certain life forms, in particular lactic acid bacteria. Consequently, fermentative bacteria are commonly employed in the food industry as starter cultures for industrial processing. On the other hand, lactic acid can be also used as food additive in the industry of edible products without the presence of lactic acid bacteria. This option can be extremely useful in various ambits. In spite of the main recommended use for this compound (intended as two optical isomers and the racemic mixture of them) as acidity regulator in many food products, several limitations should be considered in certain situations when speaking of maximum allowed amounts and the exclusive use of one isomer only. In addition, several different additives are chemically derived from lactic acid. For this reason, the application spectrum of lactic acid may be broader than the above-mentioned situation, with correlated exceptions and limitations.

Keywords Acidity regulation · Food additive · General Standard for Food Additives · Good Manufacturing Practice · Lactic acid bacteria · Lactic ester · Maximum allowed limit

Abbreviations

FAO   Food and Agriculture Organization of the United Nations
GSFA General Standard for Food Additives
GMP   Good Manufacturing Practice
LAB   Lactic acid bacteria
WHO   World Health Organization

© The Author(s) 2017
S.M. Ameen and G. Caruso, *Lactic Acid in the Food Industry*,
Chemistry of Foods, DOI 10.1007/978-3-319-58146-0_1

1.1   Introduction

The importance of lactic acid and its derivatives in the food industry is apparently linked to microbial activity of certain life forms (Chap. 5), in particular lactic acid bacteria (LAB). The production of lactic acid from hexoses is a peculiar metabolic activity, also named 'fermentation', of LAB microorganisms such as *Lactococcus, Lactobacillus, Enterococcus*, and *Streptococcus* spp. For these reasons, fermentative bacteria are commonly employed in the food industry as starter cultures for industrial processing of fermented dairy, meat, cereal, and vegetable products.

On the other hand, lactic acid can be also used as food additive in the industry of edible products without the presence of LAB fermentation. This option can be extremely useful in various ambits. The present chapter has the aim of giving a brief and reliable overview of possible uses for this important additive in the current food industry.

1.2   Lactic Acid as a Food Additive

According to the 2016 version of the General Standard for Food Additives (GSFA) CODEX STAN 192-1995 (Codex Alimentarius Commission 1995), lactic acid can be used in different ambits, although the recommended use is only related to acidity regulation. It should be considered that the name 'lactic acid' concerns three different forms: L(+), also named (S)-lactic acid; D(−) form, or (R)-lactic acid; and the racemic mixture of these isomers. L(+)-lactic acid is the biological isomer as it is naturally present in the human body; consequently, the importance of this form of lactic acid depends on the known biochemical synthesis (Chap. 2).

From a general point of view, there is only a possible recommended use (Codex Alimentarius Commission 1995) for all three additive forms, at first sight. Lactic acid is recommended as an acidity regulator for selected food categories (e.g. sterilised and UHT creams; whey protein cheese; frozen vegetables, seaweeds, and nuts and seeds; salt substitutes; formulae for special medical purposes for infants). Anyway, maximum levels are always limited by Good Manufacturing Practices (GMPs) with only one exception: 2000 g per kg when speaking of the category 13.2 'complementary foods for infants and young children'.

Actually, some interesting limitation may be observed in certain situations (Codex Alimentarius Commission 1995). Generally, the following food categories

- Infant formulae
- Follow-up formulae
- Formulae for special medical purposes for infants

show two important exceptions. The first limitation concerns the nature of formulae (complete exclusion of whole fish), while the second recommendation implies the exclusive use of the L(+)-form (as a result, dextrogire lactic acid and the racemic mixture are not considered).

The above-mentioned category 'complementary foods for infants and young children' concerns the use of lactic acid on condition that a maximum limit is allowed; in addition:

(a) The use of lactic acid includes the only levogire form, and
(b) The recommendation does not include the use of this additive in products corresponding to the 'Standard for Processed Cereal-Based Foods for Infants and Young Children' (as intended in the CODEX STAN 74-1981) at GMP (Codex Alimentarius Commission 1995).

Moreover, lactic acid is recommended in the food category 'dried pastas and noodles and like products' on condition that these foods include only 'noodles, gluten-free pasta, and pasta intended for hypoproteic diets'. Different products are not considered in this ambit.

Finally, the following two categories of vegetable foods:

- Untreated fresh vegetables, seaweeds, and nuts and seeds
- Frozen vegetables, seaweeds, and nuts and seeds can be treated with lactic acid on condition that:

(a) Interested foods comprehend only 'edible fungi and fungus products', and
(b) Lactic acid is not used in sterilised fungi exceeding 5000 mg per kg, singly or in combination with citric acid.

All these limitations should be considered carefully because the use of a food additive cannot pose safety concerns and dangers to consumers' risks. Apart from these considerations, lactic acid may be used for other food productions: as an example, it can be used in small amounts for the production of jellies and fruit syrups instead of tartaric acid (Belitz et al. 2009; Datta et al. 1995, 2006; John et al. 2007; Ueno et al. 2003). Other uses concern the contrasting action against discoloration and the enhancement of flavour properties in certain vegetables, and the improvement of egg whippability (stabilisation) by means of pH regulation. Calcium salt may be also used for milk powders with the aim of enhancing natural properties of the original milk (Belitz et al. 2009; Harada et al. 1989).

1.3 Lactic Acid Derivatives as Food Additives

Lactic acid is surely important in the food industry. On the other hand, several different additives are chemically derived from lactic acid. For this reason, the application spectrum of lactic acid may be broader than the above-mentioned situation.

The most important category of lactic acid derivatives with possible food applications is certainly the group of 'lactic and fatty acid esters of glycerol', according to the GSFA. This group of fatty esters may be used in many food productions with three main purposes (Codex Alimentarius Commission 1995):

(a) Emulsification
(b) Sequestration
(c) Stabilisation.

The use of lactic esters as emulsifying and surface active agents is well known. Mono- and diglycerides esterified with lactic acid are powerful emulsifiers. A good example can be stearyl-2-lactylate, obtained from stearic acid and lactic acid in alkaline solution (Belitz et al. 2009). Obtained lactylates are mainly represented by calcium or sodium stearyl-2-lactylate, depending on the used alkaline agent (calcium or sodium hydroxide).

Because of the chemical relationship with lactic acid, these compounds are recommended with these objectives in some of food categories already shown for lactic acid, including pasteurised cream (plain); sterilised creams; fresh pastas, noodles and similar foods; salt substitutes Interestingly, a maximum limit of 5000 g per kg is recommended when speaking of the category 13.2 'complementary foods for infants and young children'; a different and non-GMP limited values have been decided for lactic acid also in this ambit. All remaining food categories do not show similar limitations (Codex Alimentarius Commission 1995).

Actually, certain restrictions for the use of similar substances have been considered in the GSFA. Differently from lactic acid, it has to be recognised that exclusions and exceptions are more detailed when speaking of lactic esters; in addition, only three food categories—'renneted milk (plain)', 'sterilised and UHT creams, whipping and whipped creams, and reduced fat creams (plain)', and 'salt substitutes'—do not show peculiar exceptions, while peculiar and very specific indications are mentioned for all remaining categories depending on the food class. As an example, the use of similar lactic esters in the food category 09.2.4.3 ('fried fish and fish products, including mollusks, crustaceans, and echinoderms') is allowed, provided that lactic esters are used 'in breading or batter coatings only' (Codex Alimentarius Commission 1995).

References

Belitz HD, Grosch W, Schieberle P (2009) Food chemistry, 4th edn. Springer, Berlin. doi:10.1007/978-3-540-69934-7

Codex Alimentarius Commission (1995) General Standard for Food Additives (GSFA) CODEX STAN 192-1995, last amendment: 2016. Food and Agriculture Organization of the United Nations (FAO), Rome, and World Health Organization (WHO), Geneva

Datta R, Henry M (2006) Lactic acid: recent advances in products, processes and technologies—a review. J Cheml Technol Biotechnol 81(7):1119–1129. doi:10.1002/jctb.1486

Datta R, Tsai SP, Bonsignore P, Moon SH, Frank JR (1995) Technological and economic potential of poly (lactic acid) and lactic acid derivatives. FEMS Microbiol Rev 16(2–3):221–231. doi:10.1016/0168-6445(94)00055-4

Harada H, Chihara S, Suginaka Y, Suido S, Kobayashi T (1989) Process for making a calcium enriched milk. US Patent 4,840,814, 20 Jun 1989

John RP, Nampoothiri KM, Pandey A (2007) Fermentative production of lactic acid from biomass: an overview on process developments and future perspectives. Appl Microbiol Biotechnol 74 (3):524–534. doi:10.1007/s00253-006-0779-6

Ueno T, Ozawa Y, Ishikawa M, Nakanishi K, Kimura T (2003) Lactic acid production using two food processing wastes, canned pineapple syrup and grape invertase, as substrate and enzyme. Biotechnol Lett 25(7):573–577. doi:10.1023/A:1022888832278

Chapter 2
Chemistry of Lactic Acid

Abstract The importance of lactic acid in the food industry is certainly correlated with its peculiar chemical and physical properties. According to the Joint FAO/WHO Food Standards Programme, lactic acid isomers and the racemic mixture can be used as acidity regulator in certain foods with the aim of contrasting certain acid-sensitive microorganisms. As a result, the description of food-related uses of lactic acid should involve also peculiar chemical and physical features. This chapter would give a brief and accurate overview of chemical and physical features of this additive. In addition, the chemical synthetic processes for the production of the so-called milk acid are described. Finally, fermentative pathways and related industrial strategies are discussed.

Keywords Distillation · Heterolactic fermentation · Homolactic fermentation · Hydrolysis · Lactic acid · Lactic acid bacteria · Racemic mixture

Abbreviations

IUPAC   International Union of Pure and Applied Chemistry
LAB   Lactic acid bacteria
H_2SO_4 Sulphuric acid

2.1 Lactic Acid and Chemical Features. An Introduction

The importance of lactic acid in the food industry is certainly correlated with its peculiar chemical and physical properties. According to the Joint FAO/WHO Food Standards Programme–Codex Committee on Food Additives (Codex Alimentarius Commission 2013), lactic acid (intended as the two L and D isomers with the additional racemic mixture) can be used as acidity regulator in certain foods (e.g. smoked fish and smoke-flavoured fish) with the aim of contrasting certain acid-sensitive microorganisms.

© The Author(s) 2017   7
S.M. Ameen and G. Caruso, *Lactic Acid in the Food Industry*,
Chemistry of Foods, DOI 10.1007/978-3-319-58146-0_2

As a result, the description of food-related uses of lactic acid should involve also peculiar chemical and physical features. This chapter would give a brief and accurate overview of chemical and physical features. In addition, the chemical synthesis of lactic acid is described; finally, these properties are given below.

2.1.1   Basic Properties

Lactic acid, also named 'milk acid', is an organic acid with the following chemical formula: $CH_3CH(OH)CO_2H$. The official name given by the International Union of Pure and Applied Chemistry (IUPAC) is 2-hydroxypropanoic acid (Table 2.1). This important acid can be naturally produced (Martinez et al. 2013), but its importance is correlated with synthetic productions. Pure lactic acid is a colourless and hydroscopic liquid; it can be defined a weak acid because of its partial dissociation in water (Eq. 2.1) and the correlated acid dissociation constant ($K_a = 1.38 \times 10^{-4}$).

$$H_3C-CH(OH)-COOH \rightleftharpoons H^+ + H_3C-CH(OH)-COO^- \qquad (2.1)$$

Table 2.1 shows the most important data related to lactic acid.

2.1.2   Isomers

Lactic acid is a chiral compound with a carbon chain composed of a central (chiral) atom and two terminal carbon atoms (Fig. 2.1). A hydroxyl group is attached to the

Table 2.1 Physical and chemical properties of lactic acid (Igoe 2011; Mohanty et al. 2015; Vaidya et al. 2005)

Identification parameters	Description
Compound name	Lactic acid
IUPAC name	2-Hydroxypropanoic acid
Chemical formula	$C_3H_6O_3$
Molecular mass	90.08 mol g^{-1}
Appearance	A colourless and syrupy liquid; alternatively, a white to yellow solid compound
Taste	Mild acid taste
Odour	Odourless
Boiling point	122 °C
Melting point	17 °C
Specific gravity/density	1.2
K_a	1.38×10^{-4}
pK_a	3.86

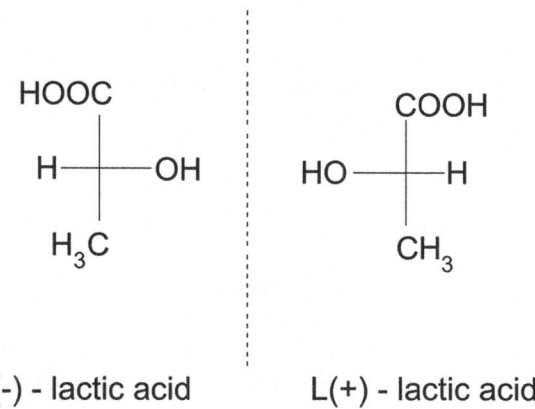

D(-) - lactic acid   L(+) - lactic acid

Fig. 2.1 Lactic acid is a chiral compound with a carbon chain composed of a central (chiral) atom and two terminal carbon atoms. A hydroxyl group is attached to the chiral carbon atom while one of the terminal carbon atoms is part of the carboxylic group and the other atom is part of the methyl group. Consequently, two optically active isomeric forms of lactic acid exist: L(+) form, also named (S)-lactic acid, and D(−) form, or (R)-lactic acid. L(+)-lactic acid is the biological isomer. BKchem version 0.13.0, 2009 (http://bkchem.zirael.org/index.html) has been used for drawing this structure

chiral carbon atom while one of the terminal carbon atoms is part of the carboxylic group and the other atom is part of the methyl group (Narayanan et al. 2004).

As a result, two optically active isomeric forms of lactic acid exist: L(+)-form, also named (S)-lactic acid, and D(−)-form, or (R)-lactic acid. Pure and anhydrous racemic mixture of lactic acid is a white crystalline solid with a low melting point. L(+)-lactic acid is the biological isomer as it is naturally present in the human body; consequently, the importance of this form of lactic acid depends of the known biochemical synthesis (Narayanan et al. 2004; Ou et al. 2011).

2.2   Synthesis of Lactic Acid

Basically, lactic acid can be produced by different chemical pathways and by microbiological synthesis. The first commercial production is ascribed to Charles E. Avery in 1881 (Carr et al. 2002; Kelkar and Mahanwar 2015).

2.2.1   Lactic Acid and Chemical Synthetic Strategies

2.2.1.1   Hydrolysis of Lactic Acid Derivatives

Lactic acid can be produced from the most part of its derivatives by means of suitable treatments (Ghaffar et al. 2014; Vaidya et al. 2005). Lactonitrile

(2-hydroxypropanenitrile, $CH_3CHOHCN$) is the most preferable of these compounds used in the chemical synthesis of lactic acid rather than other raw materials. Lactonitrile can be produced by the nucleophilic addition of hydrogen cyanide (HCN) to the liquid phase of acetaldehyde (CH_3CHO) in alkaline media under high pressure (Eq. 2.2).

$$HCN + H_3C-CHO \xrightarrow{high/pressure} H_3C-CHOH-CN \tag{2.2}$$

After recovery and distillation of the obtained impure lactonitrile (Narayanan et al. 2004), the purified compound can be treated (acid hydrolysis) by using concentrated hydrochloric acid (HCl) or concentrated sulphuric acid (H_2SO_4), with the resulting production of ammonium sulphate salt—$(NH_4)_2SO_4$—and crude lactic acid (Eq. 2.3).

$$H_3C-CHOH-CN + H_2O + \frac{1}{2}H_2SO_4 \xrightarrow{hydrolysis} H_3C-CHOH-COOH$$
$$+ \frac{1}{2}(NH_4)_2SO_4 \tag{2.3}$$

The produced (crude) lactic acid needs to be concentrated and purified. Methanol (CH_3OH) can be used with the aim of producing methyl lactate ester, $CH_3CHOHCOOCH_3$ (Eq. 2.4).

$$H_3C-CHOH-COOH + H_3-C-OH \xrightarrow{esterification} H_3C-CHOH-COOCH_3 + H_2O \tag{2.4}$$

Methyl lactate ester is subsequently collected, purified by distillation, and hydrolysed in acidic aqueous solution to lactic acid, while methanol can be recycled in the same process (Eq. 2.5). The resulting product is a racemic mixture of lactic acid (Narayanan et al. 2004).

$$H_3C-CHOH-COOCH_3 + H_2O \xrightarrow{hydrolysis} H_3C-CHOH-COOH + H_3C-OH \tag{2.5}$$

2.2.1.2   Nitric Acid Oxidation of Propane

Another pathway for the chemical synthesis of lactic acid concerns the use of propene ($CH_3CH_2CH_3$). This alkene is oxidised to α-nitropropionic acid (Vaidya et al. 2005) by using nitric acid (HNO_3) in presence of oxygen (Eq. 2.6). Subsequently, the obtained acid has to be converted into lactic acid by hydrolysis (Eq. 2.7).

$$H_3C-CH_2-CH_3 + HNO_3 \xrightarrow{\text{oxydation/O}_2} H_3C-CH-NO_2-COOH \qquad (2.6)$$

$$H_3C-CH-NO_2-COOH + H_2O \xrightarrow{\text{hydrolysis}} H_3C-CHOH-COOH + HNO_3 \quad (2.7)$$

2.2.2 Lactic Acid Fermentation

The mass production of lactic acid by using fermentation became widely used after the discovery of *Lactobacillus sp.* by the French chemist Loius Pasteur in 1856. *Lactobacillus* bacteria are able to produce lactic acid from carbohydrates such as glucose and lactose, and they are even living in our gastrointestinal system (Carr et al. 2002). Fermentation is a biochemical process in which carbohydrate molecules, e.g. glucose, are converted into energy, lactate, and other by-products depending on the type of microorganism involved in the fermentation process (John et al. 2007). For these reasons, there are two basic fermentation processes (Sect. 5.1): the homolactic fermentation (prevailing product: lactic acid), and a heterolactic fermentation (the final product is a mixture containing mainly lactic acid, other organic acids, and ethyl alcohol). Both mechanisms are shown in Figs. 2.2 and 2.3 (Fugelsang and Edwards 2007). Other fermentation types can occur depending on the fermentation raw materials and conditions (Ikushima et al. 2009; Wasewar et al. 2002; Wee et al. 2005).

2.2.2.1 Solid-state Fermentation

In relation to solid-state fermentation, microbial growth and fermentation take place at the surface of solid substrates such as wheat, soya bean, and cheeses. Such substrates are more convenient for a large number of filamentous fungi and a few bacteria (Chisti 1999; Jelen 2003; Kim et al. 2006).

2.2.2.2 Submerged Fermentation

In this fermentation, the substrate for microbial growth is the liquid solution placed in large tanks called 'fermenters' or 'bioreactors' (Chisti 1999; Wee et al. 2006). Submerged fermentation can be subdivided in three categories:

- Batch fermentation. Substrates and required raw materials for fermentation and the desired microbial growth are placed into a bioreactor; incubation is allowed to proceed on condition that operating parameters such as pH and thermal values are defined. During fermentation, nothing is added except oxygen in case of aerobic microorganisms. After each process, the product is collected and the

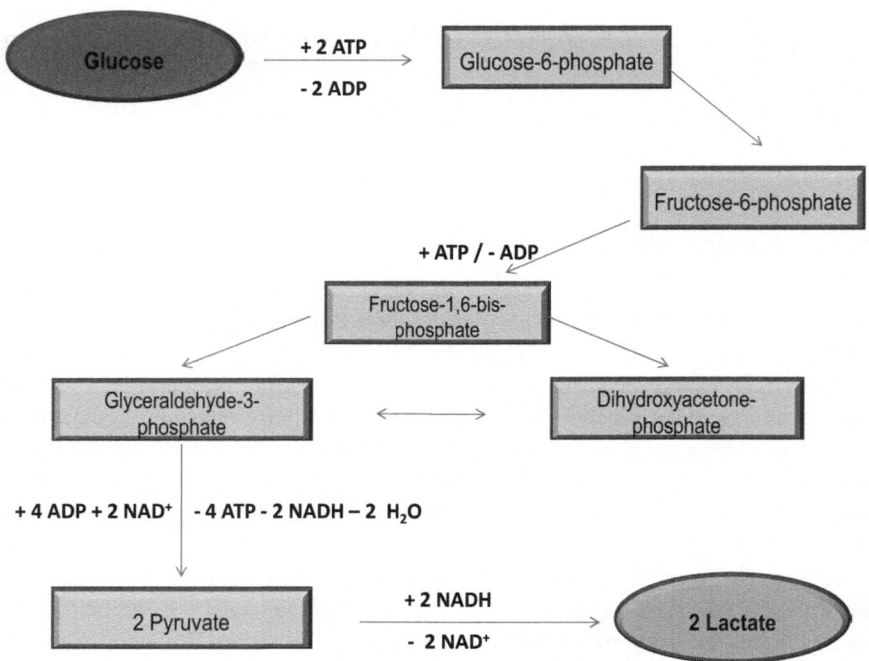

Fig. 2.2 LAB and lactic acid synthesis: the homofermentative mechanism (Embden–Meyerhof–Parnas pathway), from one single glucose molecule to two lactate units. *ATP* means: adenosine triphosphate; *ADP* is for: adenosine diphosphate; *NAD⁺* and *NADH* correspond to nicotinamide adenine dinucleotide reduced and non-reduced forms

fermenter is cleaned; then, another batch can be prepared and the process may restart
- Fed-batch fermentation. Substrates and raw materials are added in small amounts during the fermentation process. Both batch and fed-batch procedures are considered as 'closed' fermentation systems, differently from 'open' systems such as continuous fermentation (Portno 1968)
- Continuous fermentation. The addition of substrates and raw materials is performed continuously during the process. Consequently, continuous fermentation is considered as an open system: The introduction of new raw materials is allowed, differently from 'closed' systems (Portno 1968).

The highest concentration of lactic acid is normally obtained in batch and fed-batch cultures (discontinuous processes), whereas the highest productivity is observed in continuous fermentation processes because of longer temporal periods (working cycles).

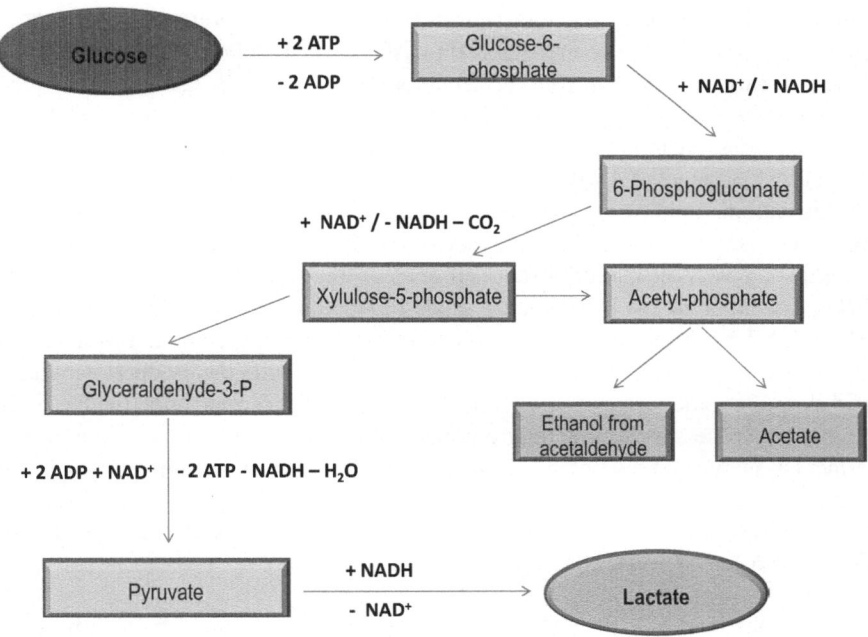

Fig. 2.3 LAB and lactic acid synthesis: the heterofermentative mechanism, from one single glucose molecule to one lactate unit. *ATP* means: adenosine triphosphate; *ADP* is for: adenosine diphosphate; *NAD⁺* and *NADH* correspond to nicotinamide adenine dinucleotide reduced and non-reduced forms. This mechanism gives also ethanol (from acetaldehyde) and acetate

2.2.2.3   Anaerobic Fermentation

This fermentation process involves anaerobic microorganisms; the air into fermenters is replaced by carbon dioxide, hydrogen, nitrogen, or a suitable mixture of these gases.

2.2.2.4   Aerobic Fermentation

Fermentation process can be also carried out in presence of aerobic microorganisms under aerobic conditions. Raw materials and conditions used for lactic acid production, e.g. purity, pH, and temperature, are critical parameters for the further purification of obtained and impure lactic acid (Krishna et al. 2012). Monosaccharides (e.g. glucose) and disaccharides (e.g. sucrose, maltose, and lactose) are common substrates for this process (Lunelli et al. 2010, 2011). Monosaccharide and disaccharide are end products of starch hydrolysis by application of enzymes such as glucoamylases and α-amylases or by chemical hydrolysis, since most microorganisms cannot utilise polysaccharides such as starch without hydrolysis. The choice of substrate and other conditions are dependent on microorganisms used in fermentation (Fukushima et al. 2004).

Life forms used in industrial fermentation can be bacteria such as *Escherichia coli* and *Lactobacillus* spp., or fungal organisms such as *Rhizopus* spp. In relation to food industries, bacteria involved in fermentation are named as 'lactic acid bacteria' (LAB) such as genera *Lactobacillus, Streptococcus, Leuconostoc*, and *Pediococcus*. In this ambit, conditions such as temperature, pH, aeration, and agitation are important parameters and they can vary depending on the type of LAB; so, these conditions have to be carefully set (Carr et al. 2002; Chooklin et al. 2011; Ge et al. 2010; Coelho et al. 2011; Narayanan et al. 2004; Romani et al. 2008; Secchi et al. 2012). LAB reach the maximum productivity only within specific temperature and pH ranges, while the fermentation process is associated with the production of lactic acid as well as other organic acids which lower the pH of fermentation media (or the broth). It is necessary to maintain optimum pH values at a constant value during fermentation by addition of alkali such as hydroxides or calcium carbonate, or ammonia. Calcium hydroxide, $Ca(OH)_2$, can react with carbohydrates such as glucose (Vaidya et al. 2005) with the production of calcium lactate, $(H_3C–CHOH–COO^-)_2\ Ca^{2+}$, and water (Eq. 2.8).

$$C_6H_{12}O_6 + Ca(OH)_2 \xrightarrow{\text{fermentation/enzymes}} (H_3C-CHOH-COO^-)Ca^{2+} + 2H_2O$$

$$(2.8)$$

Calcium lactate has to be filtered and separated from the obtained aqueous solution, treated with H_2SO_4 to be hydrolysed, and turned into lactic acid and calcium sulphate (Eq. 2.9).

$$2(H_3C-CHOH-COO^-)Ca^{2+} + H_2SO_4 \xrightarrow{\text{hydrolysis}} 2H_3C-CHOH-COOH + CaSO_4$$

$$(2.9)$$

Obtained lactic acid is separated by filtration of calcium sulphate; subsequently, purification and esterification with methanol are needed to obtain methyl lactate which undergoes hydrolysis to pure lactic acid (Narayanan et al. 2004; Vaidya et al. 2005) as shown in Eqs. 2.4 and 2.5. The output of fermentation is an aqueous lactic acid solution which is subsequently concentrated by evaporation (Komesu et al. 2013; Martins et al. 2012, 2013).

LAB utilise either the well-known Embden–Meyerhoff–Parnas pathway of glucose metabolism to obtain lactic acid as the major end product, or use pathways of pentose metabolism resulting in the formation of lactic acid plus other products such as acetic acid, ethanol, and carbon dioxide.

A limited number of non-LAB microorganisms are capable to produce larger amounts of lactic acid from common carbon sources if compared with LAB. The best known of these life forms is *Rhizopus oryzae* which can be used for commercial lactic acid production since it can convert several sugars. The average lactic acid yield seems to be approximately around 93.8 g per l, while other LAB (*E. faecalis*) are reported to produce higher amounts and different Lactobacilli can produce lactic acid in the range 21.8–90.0 with average amounts of 60.3 g per l. Anyway, these microorganisms may be really different when speaking of productivity values in terms of grams per litre in one single hour (Abdel-Rahman et al. 2011; Wee et al. 2006).

Unlike the chemical synthesis, fermentation processes for the production of lactic acid can give selectively one of the two lactic acid stereoisomers or their racemic mixture depending on the bacteria species selection (Martinez et al. 2013). Industrial production of lactic acid, especially where pure optical isomers are required, is presently carried out predominantly by fermentation processes. In summary, the production process can be subdivided into two steps:

1. The real production of lactic acid by fermentation of a carbohydrate source.
2. The downstream processing of the fermentation broth to obtain pure lactic acid.

References

Abdel-Rahman MA, Tashiro Y, Sonomoto K (2011) Lactic acid production from lignocellulose-derived sugars using lactic acid bacteria: overview and limits. J Biotechnol 156(4):286–301. doi:10.1016/j.jbiotec.2011.06.017

Carr FJ, Chill D, Maida N (2002) The lactic acid bacteria: a literature survey. Crit Rev Microbiol 28(4):281–370. doi:10.1080/1040-840291046759

Chisti Y (1999) Fermentation (industrial): basic considerations. In: Robinson R, Batt C, Patel P (eds) Encyclopedia of food microbiology. Academic Press, London, pp 663–674

Chooklin S, Kaewsichan L, Kaewsichan L (2011) Potential use of lactobacillus casei TISTR 1500 for the bioconversion from palmyra sap and oil palm sap to lactic acid. Electron J Biotechnol 14(5):10. doi:10.2225/vol14-issue5-fulltext-11

Codex Alimentarius Commission (2013) Endorsement and/or revision of maximum levels for food additives and processing aids in codex standards. Joint FAO/WHO Food Standards Programme —Codex Committee on Food Additives, Agenda Item 4a CX/FA 13/45/4, Forty-fifth Session, Beijing, China, 18–22 March 2013. Available ftp://fao.org/codex/meetings/ccfa/ccfa45/fa45_04e.pdf. Accessed 30 Jan 2017

Coelho LF, de Lima CJB, Bernardo MP, Contiero J (2011) D(-)-lactic acid production by Leuconostoc mesenteroides B512 using different carbon and nitrogen sources. Appl Biochem Biotechnol 164(7):1160–1171. doi:10.1007/s12010-011-9202-6

Fugelsang KC, Edwards CG (2007) Wine microbiology practical applications and procedures. Springer, New York

Fukushima K, Sogo K, Miura S, Kimura Y (2004) Production of D-lactic acid by bacterial fermentation of rice starch. Macromol Biosci 4(11):1021–1027. doi:10.1002/mabi.200400080

Ge XY, Qian H, Zhang WG (2010) Enhancement of L-lactic acid production in *Lactobacillus casei* from Jerusalem Artichoke Tubers by Kinetic optimization and citrate metabolism. J Microbiol Biotechnol 20(1):101–109

Igoe RS (2011) Dictionary of food ingredients, 5th edn. Springer Science+Business Media, New York

Ikushima S, Fujii T, Kobayashi O, Yoshida S, Yoshida A (2009) Genetic engineering of Candida utilis yeast for efficient production of L-lactic acid. Biosci Biotechnol Biochem 73(8): 1818–1824. doi:10.1271/bbb.90186

Jelen P (2003) Whey processing, utilization and products. In: Roginski H, Fuquay JW, Fox PF (eds) Encyclopedia of dairy sciences. Academic Press, London, pp 2739–2751

John RP, Nampoothiri KM, Pandey A (2007) Fermentative production of lactic acid from biomass: an overview on process developments and future perspectives. Appl Microbiol Biotechnol 74(3):524–534. doi:10.1007/s00253-006-0779-6

Kelkar ST, Mahanwar PA (2015) Production of lactic acid from Tamarind Kernel by *Lactobacillus Casei*. Int J Technol Enhanc Emerg Eng Res 3(5):25–31

Kim HO, Wee YJ, Kim JN, Yun JS, Ryu HW (2006) Production of lactic acid from cheese whey by batch and repeated batch cultures of *Lactobacillus* sp RKY2. Appl Microbiol Biotechnol 131(1–3):694–704. doi:10.1385/ABAB:131:1:694

Komesu A, Martins PF, Lunelli BH, Morita AT, Coutinho PLA, Maciel Filho R, Wolf Maciel MR (2013) Lactic acid purification by hybrid short path evaporation. Chem Eng Trans 32: 2017–2022. doi:10.3303/CET1332337

Krishna G, Rangaiah GP, Lakshminarayanan S (2012) Modeling and analysis of intensified processes for economic recovery of high-grade lactic acid. In: Proceedings of the 22nd European symposium on computer aided process engineering, England, London, 17–20 June

Lunelli BH, Andrade RR, Atala DIP, Wolf Maciel MR, Maugeri Filho F, Maciel Filho R (2010) Production of lactic acid from sucrose: strain selection, fermentation and kinetic modeling. Appl Biochem Biotechnol 161(1–8):227–237. doi:10.1007/s12010-009-8828-0

Lunelli BH Morais ER, Wolf Maciel MR, Maciel Filho R (2011) Process intensification for ethyl lactate production using reactive distillation. In: Proceedings of the 10th international conference on chemical and process engineering, 8–11 May 2011, Vol 24, Florence, Italy, doi:10.3303/CET1124138

Martinez FAC, Balciunas EM, Salgado JM, Gonzalez JMD, Converti A, Oliveira RPS (2013) Lactic acid properties, applications and production: a review. Trends Food Sci Tech 30(1): 70–83. doi:10.1016/j.tifs.2012.11.007

Martins PF, Carmona C, Martinez EL, Sbaite P, Maciel Filho R, Wolf Maciel MR (2012) Evaluation of methyl chavicol concentration by different evaporation processes using central composite experimental design. Sep Purif Technol 98:464–471. doi:10.1016/j.seppur.2012.08.009

Martins PF, Medeiros HHR, Sbaite P, Wolf Maciel MR (2013) Enrichment of oxyterpenes from orange oil by short path evaporation. Sep Purif Technol 116:385–390. doi:10.1016/j.seppur.2013.06.011

Mohanty JN, Das PK, Nanda S, Nayak P, Pradhan P (2015) Comparative analysis of crude and pure lactic acid produced by *Lactobacillus fermentum* and its inhibitory effects on spoilage bacteria. Pharm Innov J (3)11:38–42. Available http://www.thepharmajournal.com/archives/2015/vol3issue11/PartA/10.1.pdf. Accessed 10 Feb 2017

Narayanan N, Roychoudhury PK, Srivastava A (2004) L(+) lactic acid fermentation and its product polymerization. J Biotechnol 7(2):167–179

Ou MS, Ingram LO, Shanmugam KT (2011) L(+)-lactic acid production from non-food carbohydrates by thermotolerant *Bacillus coagulans*. J Ind Microbiol Biotechnol 38(5): 599–605. doi:10.1007/s10295-010-0796-4

Portno AD (1968) Continuous fermentation of Brewer's wort. J Inst Brew 74(1):55–63. doi:10.1002/j.2050-0416.1968.tb03096.x

Romani A, Yanez R, Garrote G, Alonso JL (2008) SSF production of lactic acid from cellulosic biosludges. Bioresour Technol 99(10):4247–4254

Secchi N, Giunta D, Pretti L, Garcia MR, Roggio T, Mannazzu I, Catzeddu P (2012) Bioconversion of ovine scotta into lactic acid with pure and mixed cultures of lactic acid bacteria. J Ind Microbiol Biotechnol 39(1):175–181. doi:10.1007/s10295-011-1013-9

Tayyba Ghaffar A, Muhammad Irshad A, Zahid Anwar A, Tahir Aqil B, Zubia Zulifqar A, Asma Tariq A, Muhammad Kamran A, Nudrat Ehsan A, Mehmood Sajid (2014) Recent trends in lactic acid biotechnology: a brief review on production to purification. J Radiat Res Appl Sci 7(2):222–229. doi:10.1016/j.jrras.2014.03.002

Vaidya AN, Pandey RA, Mudliar S, Kumar MS, Chakrabart T, Devotta S (2005) Production and recovery of lactic acid for polylactide—an overview. Crit Rev Environ Sci Technol 35(5): 429–467. doi:10.1080/10643380590966181

Wasewar KL, Heesink ABM, Versteeg GF, Pangarkar VG (2002) Reactive extraction of lactic acid using alamine 336 in MIBK: equilibria and kinetics. J Biotechnol 97(1):59–68. doi:10.1016/S0168-1656(02)00057-3

Wee YJ, Yun JS, Lee YY, Zeng AP, Ryu HW (2005) Recovery of lactic acid by repeated batch electrodialysis and lactic acid production using electrodialysis wastewater. J Biosci Bioeng 99 (2):104–108. doi:10.1263/jbb.99.104

Wee YJ, Kim JN, Ryu HW (2006) Biotechnological production of lactic acid and its recent applications. Food Technol Biotechnol 44:163–172

Chapter 3
Regulatory Importance of Lactic Acid in the Food and Beverage Sector

Abstract Since the discovery of lactic acid, this compound has been considered for many significant uses and applications in food, cosmetic, pharmaceutical, and chemical industries. Lactic acid occurs naturally in many edible products and also an important ingredient in the food industry. Also, lactic acid is non-toxic and consequently recognised and classified as 'Generally Recognized as Safe' substance by the United States Food and Drug Administration for wide use as an additive in the food industry. The notable range of legally allowed food applications depends on the demonstrated antimicrobial action against pathogens and the concomitant shelf life extension. In addition, this acid can serve as a flavouring substance in many food products such as pickles and fermented milk. This chapter discusses some practical and legally allowed examples of lactic acid uses in the production of fermented vegetables, cheeses, fermented milks, sourdough, fermented meats, and wines.

Keywords Fermentation · Flavouring agent · Lactic acid · Lactobacilli · Preservation · Shelf life extension · Streptococci

Abbreviations

LAB   Lactic acid bacteria

3.1 Introduction

Since the discovery of lactic acid, this compound has been considered for many significant uses in food, cosmetic, pharmaceutical, and chemical industries (Chooklin et al. 2011; Newton 2007; Vaclavik and Christian 2014; Vieira 1996). Lactic acid occurs naturally in many edible products and also an important ingredient in the food industry. Also, lactic acid is non-toxic and consequently recognised and classified as 'Generally Recognized as Safe' substance by the United States Food and Drug Administration for wide use as an additive in the food

© The Author(s) 2017
S.M. Ameen and G. Caruso, *Lactic Acid in the Food Industry*,
Chemistry of Foods, DOI 10.1007/978-3-319-58146-0_3

industry. The notable range of possible food applications, including pH regulation and flavouring properties, is correlated with its chemical and physical properties (Sect. 2.1).

The direct addition of lactic acid during the process is not strictly required in many food productions. Although lactic acid is named 'milk acid' (Sect. 2.1.1), this acid is not derived from milk but formed during fermentation in many nondairy products such as beer, fermented fruits and vegetables, fermented sausages, and sourdough. In detail, lactic acid produced as the end product of carbohydrate fermentation is obtained by means of the action of lactic acid bacteria (LAB). Lactic acid is produced in the form of L(+)- or D(−)-lactic acid or as the racemic mixture of these isomers, depending on microorganisms involved in fermentation (Belitz et al. 2009; Damodaran et al. 2008; Manley 2000). Some *Lactobacilli* sp. Bacteria produce L(+)-lactic acid form and *Lactobacillus plantarum* produce both L(+)- and D(−)-forms. L(+)-lactic acid is naturally synthesised in biological systems, and hence its use as an acidulant does not introduce a foreign element into the body (Nikkilä et al. 2009).

3.2   Preservation and Flavouring

Lactic acid is used in a wide range of food industry sectors such as bakery products, beverages, fermented meat, confectionery, dairy products, salads, dressings, and ready meals. This additive is very important in the food preservation sector due to its antimicrobial action against pathogens beside microflora regulation which increases the shelf life of edible products (Belitz et al. 2009; Rodríguez et al. 2002). Lactic acid fermentation is considered as one of the important approaches used for food processing and preserving throughout the world, at present. This acid usually serves as a preservative agent and as a flavouring substance in many food products at the same time. Beside its preservative action, its mild acidic taste (souring) explains well its role as flavouring agent, especially when speaking of fermented food products such as pickles and fermented milk. Along with holding at suitable temperatures, certain dairy products and fermented vegetables are preserved with lactic acid which is released during microbial activity (Belitz et al. 2009; Steinkraus 1983). Some legally allowed examples of the use of lactic acid in the food preservation sector are shown in the next section.

3.2.1   Lactic Acid in Fermented Vegetables

Lactic acid can be used as an acidifying agent by direct addition to vegetables and fruits with the purpose of reducing pH values below four in order to control microbial growth (Guizani and Mothershaw 2007; Steinkraus 1983).

Another approach to pH lowering is the action of LAB in vegetables, with consequent lactic acid production: In these conditions, fermented vegetables under suitable environment for LAB growth show the typical acid flavour. Vegetables contain water up to 95% or greater values. Should salt be added to raw vegetables (a similar approach can be performed with immersion of raw foods in concentrated salt solutions), the vegetable bulk would absorb salt from the surrounding environment until the equilibrium between concentrations on the external and the inner sides of vegetables is reached; the fermentation can take place in these conditions. Fermentation can either be achieved by addition of LAB starter culture under controlled conditions.

Various vegetables can be treated in this way (Guizani and Mothershaw 2007; Steinkraus 1983):

- Pickles. Pickling of vegetables is commonly used in vegetable preservation. A wide variety of vegetables can be pickled, e.g. carrots, cucumbers, turnips, cabbage, olives, onions, peppers, lime, etc. Pickling can be done with using LAB cultures such as *Lactobacillus* sp. During fermentation, the produced lactic acid gives the mild acidic taste.
- Sauerkraut. Like other preserved vegetables, cabbage can be preserved with lactic acid fermentation. The most common LAB types involved in the fermentation are *Leuconostoc mesenteroides*, *Lactobacillus cucumeris*, and *L. pentoaceticus* (the last lactobacillaceae are LAB life forms). *L. mesenteroides* are heterofermentative bacteria, and they are able to produce both lactic acid and acetic acid, with the additional alcohol production, in a wide thermal range.

3.2.2 Lactic Acid and Cheeses. Making and Ripening

The addition of lactic acid, rennet, or proteolytic enzymes (rennin) to milk leads to the coagulation and precipitation of a specific group of milk proteins, 'casein', which can be later separated with suitable treatments with the production of curds. Curd contains, beside coagulated casein, fat, other milk proteins (whey protein is not precipitated at acidic pH values), minerals, vitamins, and carbohydrates (lactose). This curd is the main substrate for different cheese types (Vaclavik and Christian 2014).

Since rennin requires slightly acidic pH values, lactic acid turns the medium (milk) into an acidic liquid with the consequent catalysing action of proteolytic enzymes such as rennin; as a result, biochemical reactions and milk protein coagulation can take place. Lactic acid may be added to the milk in calculated concentrations, or it may be produced by LAB. Microbial enzymes can be either responsible for coagulation and curd formation. Residual lactose which remains after casein coagulation is fermented under the action of microbial activity during the 'ripening' step. Ripening is responsible for typical cheese texture and flavour.

Cheese produced in the presence of microbial activity (bacteria, fungi, or yeasts) includes cheddar, edam, and roquefort types. On the other hand, cheese productions that can be performed without the need of ripening process include cottage, cream cheese, and ricotta cheese products.

3.2.3 Lactic Acid and Fermented Milks

Fermented milk products are used in different countries in the daily nutritional diet (Holzapfel et al. 2001; Panesar 2011). Lactose is the main compounds among milk carbohydrates, and it is soluble in water (with relation to milk, lactose is found in small amounts, e.g. 5%). Historically, fermented milks were prepared by simple slow fermentation of milk (lactose is converted into lactic acid) under the action of microflora inherently present in the initial milk. At present, a variable number of different fermented milk products are produced using some microbiological approaches under controlled conditions (Widyastuti et al. 2014).

The addition of bacterial culture to milk, such as *Lactobacilli* and *Streptococci*, induces fermentation of lactose into lactic acid with pH lowering. Starter cultures differ for each product. The most popular genre of bacteria used in dairy products is *Lactobacillus* ssp. Fermentation products can include both alcohol and lactic acids, or lactic acid only, depending on the type of used bacteria (Kuipers et al. 2000). Bacterial growth is thermally favoured depending on the specific bacterial culture. For example, *Streptococcus thermophilus* grows optimally in the range 45–47 °C, while the optimal growth of *Lactobacillus bulgaricus* is at 37 °C.

Besides casein coagulation and the consequent yogurt-like texture, the presence of lactic acid improves flavour and product variety and suppresses the growth of pathogens and spoilage microorganisms (Avonts et al. 2004; Chen and Hoover 2003). The two following products are examples of commonly fermented milk products.

- Yogurt. Yogurt is one of fermented milk-based products obtained by treatment of pasteurised milk with characteristic LAB starter cultures, mainly *L. bulgaricus* and *S. thermophilus*, under controlled environment. The joint action of both bacteria can accelerate the process of lactic acid production. Yogurt may be made using whole, low fat, or skimmed milk. Lactic acid improves yogurt consistency giving its characteristic texture and sour taste.
- Kefir fermentation. Differently from yogurt, kefir production requires starter cultures of yeasts (e.g. *Saccharomyces kefir* and *Torula kefir*) and bacteria (e.g. *L. caucasicus*) at the same time (room temperature). During fermentation, lactic acid is released with some alcohol and carbon dioxide (Newton 2007) giving the peculiar fizzy looking. Kefir taste is slightly alcoholic with the consistency of thin yogurt. The final product is white to yellowish in colour with a yeasty aroma. Should yeast be absent in the production process, kefir would not show the presence of carbon dioxide (absence of carbonation) and alcohol;

consequently, kefir would be very similar to yogurt. The amount of produced lactic acid during fermentation determines the flavour and consistency of kefir. This product contains carbon dioxide with the consequent refreshing appearance; it can be flavoured with fruits, honey, and any other sweeteners and flavours.

3.2.4 Lactic Acid and Sourdough

In baking, the mixture of soda (mainly sodium bicarbonate) and acid or salt is an important factor when speaking of 'dough leavening'. Chemical reactions between these components will release carbon dioxide (leavening agent) either before or during dough baking. In addition, the amount and size of produced gas bubbles gives the peculiar and desired texture. The importance of lactic acid here is correlated to dough leavening, but also to the control of enzymatic activity in the dough, enhancing dough taste and aroma. Moreover, spoilage microorganisms and moulds are inhibited with the consequent dough preservation and shelf life extension.

With reference to sourdough making, the main acidulant agent is lactic acid (Manley 2000; Vaclavik and Christian 2014). Carbon dioxide is produced either by reaction of soda with acid (added directly or sometimes as soured milk) or through the metabolism of carbohydrates by yeast cells or bacteria (e.g. *L. sanfrancisco*) during the fermentation process. The most common yeast strain used for bread making is *Saccharomyces exiguous*, also named 'non-baker's yeast'. Starter cultures used for fermentation can be homofermentative or heterofermentative life forms, depending also on the type of flour used for baking. The production of rye bread is normally performed with high lactic acid sourdough and heterofermentative cultures, while sourdough used for mixed wheat bread needs higher amounts of acetic acid.

3.2.5 Lactic Acid and Fermented Meat

Lactic acid has preservative effects on the meat. During the processing of fermented sausage, bacterial fermentation is allowed with suitable starter culture. Lactic acid is produced with pH adjustment around five; preservation is obtained in this way with additional features such as the peculiar flavour of fermented sausages.

3.2.6 Lactic Acid and Winemaking

Two fermentation processes are involved in winemaking. Special winemaking yeast strains and starter culture are added to the grape must in the first fermentation; yeast can metabolise all the bioavailable sugar in the medium releasing carbon dioxide and alcohol. This fermentation is followed by other production stages including the second (malolactic) fermentation. Malic, citric, and tartaric acids are naturally present in grape juice and characterised by their sharp acid taste. LAB starter culture is added in the malolactic fermentation with the aim of turning the naturally present malic acid into lactic acid with consequent low wine acidity and more acceptable tastes (Theisen and Montalbano 2008).

3.3 pH Regulation

Microorganisms are able to release many enzymes during their biological activity and growth. These enzymes are protein molecules with amphoteric nature: They can react as acid or alkaline compounds. At certain pH, enzymes become inactive and enzymatic reactions are necessarily stopped. The addition of synthetic food-grade acids or alkali (or naturally present substances) does not change or maintain only the initial pH of foods; in fact, microbial activity is notably inhibited by means of the modification of suitable pH for enzymatic activity. Moreover, pH lowering to certain levels can denature proteins of microbial cell walls leading to cellular death.

Lactic acid is a very common acidity regulator, and it can modify acidity in products such as cheese and fermented sausages with good results. This compound is normally used in industries such as confectionery to set the food medium at a suitable pH during processing steps. Lactic acid salts, such as calcium and sodium lactates, are very soluble; consequently, they may be also used for buffering purposes instead of other acids (Liu 2003).

References

Avonts L, Van Uytven E, De Vuyst L (2004) Cell growth and bacteriocin production of probiotic *Lactobacillus* strains in different media. Int Dairy J 14(11):947–955. doi:10.1016/j.idairyj. 2004.04.003

Belitz HD, Grosch W, Schieberle P (2009) Food chemistry, 4th edn. Springer, Berlin. doi:10.1007/ 978-3-540-69934-7

Chen H, Hoover DG (2003) Bacteriocins and their food applications. Compr Rev Food Sci Food Saf 2(3):82–100. doi:10.1111/j.1541-4337.2003.tb00016.x

Chooklin S, Kaewsichan L, Kaewsichan L (2011) Potential use of *Lactobacillus* casei TISTR 1500 for the bioconversion from palmyra sap and oil palm sap to lactic acid. Electron J Biotechnol 14(5):10. doi:10.2225/vol14-issue5-fulltext-11

Damodaran S, Parkin LK, Fennema RO (eds) (2008) Femmema's food chemistry, 4th edn. CRC Press, Boca Raton

Guizani N, Mothershaw A (2007) Fermentation as a method for food preservation. In: Shafiur Rahman M (ed) Handbook of food preservation, 2nd edn. CRC Press, Boca Raton

Holzapfel WH, Haberer P, Geisen R, Bjorkroth J, Schillinger U (2001) Taxonomy and important features of probiotic microorganisms in food nutrition. Am J Clin Nutr 73(Suppl 2):365S–373S

Kuipers OP, Buist G, Kok J (2000) Current strategies for improving food bacteria. Res Microbiol 151(10):815–822. doi:10.1016/S0923-2508(00)01147-5

Liu SQ (2003) Practical implications of lactate and pyruvate metabolism by lactic acid bacteria in food and beverage fermentations. Int J Food Microbiol 83(2):115–131. doi:10.1016/S0168-1605(02)00366-5

Manley D (2000) Technology of biscuits, crackers and cookies, 3rd edn. Woodhead Publishing Ltd, Cambridge, and CRC Press LLC, Boca Raton

Newton ED (2007) Food chemistry. Facts on File, Inc., New York

Nikkilä KK, Mervi H, Matti L, Airi P (2009) Metabolic engineering of *Lactobacillus helvicticus* CNRZ32 for production of pure, *L*(+) lactic acid. Appl Environ Microbiol 66(9):3835–3841

Panesar SP (2011) Fermented dairy products: starter cultures and potential nutritional benefits. Food Nutr Sci 2(1):47–51. doi:10.4236/fns.2011.21006

Rodríguez JM, Martínez MI, Kok J (2002) Pediocin PA-1, a wide-spectrum bacteriocin from lactic acid bacteria. Crit Rev Food Sci Nutr 42(2):91–121. doi:10.1080/10408690290825475

Steinkraus KH (1983) Lactic acid fermentation in the production of foods from vegetables, cereals and legumes. Antonie Van Leeuwenhoek 49(3):337–348

Theisen M, Montalbano P (eds) (2008) Guide to red winemaking. MoreFlavor! Inc., Pittsburg

Vaclavik VA, Christian EW (2014) Essentials of food science, 4th edn. Springer, New York

Vieira RE (1996) Elementary food science, 4th edn. Chapman & Hall, New York

Widyastuti Y, Rohmatussolihat Febrisiantosa A (2014) The role of lactic acid bacteria in milk fermentation. Food Nutr Sci 5:435–442. doi:10.4236/fns.2014.54051

Chapter 4
Lactic Acid in the Food Matrix: Analytical Methods

Abstract The aim of this chapter is to give a quick and sufficiently comprehensible overview of the current analytical methods for the quantitative and qualitative determination of lactic acid in foods. The importance of lactic acid in the world of food production is well known at present. At the same time, food technologists need accurate, fast, and reliable analytical methods for the determination of naturally present and/or added lactic acid (and related derivatives) in food products and in raw materials. The same thing can be affirmed when speaking of lactic acid as a normal food additive. Naturally, regulatory requirements represent an important demand for adequate and reliable analytical procedures. As a result, the 'milk acid' should be detected in foods in a reliable way, on condition that the food matrix and the desired function of lactic acid are well considered. For these reasons, this chapter discusses the practical advantages of several analytical procedures, including spectrophotometry, gas chromatography, capillary electrophoresis, and high-performance liquid chromatography, in relation to selected food categories.

Keywords Biosensors · Capillary electrophoresis · Gas chromatography · High-performance liquid chromatography · Lactic acid · Potentiometric titration · Spectrophotometry

Abbreviations

GC   Gas chromatography
GSFA   General Standard for Food Additives
HPLC   High-performance liquid chromatography
FAO   Food and Agriculture Organization of the United Nations
WHO   World Health Organization

© The Author(s) 2017
S.M. Ameen and G. Caruso, *Lactic Acid in the Food Industry*,
Chemistry of Foods, DOI 10.1007/978-3-319-58146-0_4

4.1 Introduction

The importance of lactic acid in the world of food production is well known at present (Chap. 1). At the same time, food technologists need accurate, fast, and reliable analytical methods for the determination of naturally present and/or added lactic acid (and related derivatives) in food products and in raw materials. The same thing can be affirmed when speaking of lactic acid as a normal food additive.

This molecule is widely used in different fields such as food, cosmetic, pharmaceutical, and chemical industries that are accustomed to consider lactic acid (Sect. 3.1). In addition, it is non-toxic and consequently recognised and classified as 'Generally Recognized as Safe' substance by the United States Food and Drug Administration for wide use as an additive in the food industry. Chemical and physical properties of lactic acid can explain the wide use of this additive in the current industry on the one side and the regulatory importance on the other hand.

Naturally, regulatory requirements represent an important demand for adequate and reliable analytical procedures concerning the determination of lactic acid. Two factors should be considered:

(a) This additive is very important in the food preservation sector (bakery products, dressings, ready meals, confectionery, dairy products, etc.) due to its antimicrobial action against pathogens and potential durability enhancement
(b) In addition, lactic acid usually serves in many food products as a preservative agent and as a flavouring substance at the same time.

As a consequence, the 'milk acid' (Sect. 2.1.1) should be detected in foods in a reliable way, but two basic variables have to be considered:

1. The food matrix, and
2. The desired function of lactic acid, as an additive or a naturally present compound in raw materials, into the final product.

This chapter would give a quick and sufficiently comprehensible overview of the current analytical methods for the quantitative and qualitative determination of lactic acid in foods.

4.2 Lactic Acid in Foods: Analytical Procedures

The presentation of the most useful analytical procedures concerning the determination of lactic acid in foods cannot be shown without a preliminary discrimination based on the peculiar food matrix.

In general, the following food categories may be considered at first sight when speaking of edible products for human consumption containing naturally present or added lactic acid (FAO 1986):

- Beverages and beverage ingredients (non-carbonated water-based flavoured drinks, including lactic acid beverages, Nigerian *burukntu*, other beverages obtained from fermentation of starchy-based foods)
- Canned foods (actually, this subcategory of preserved foods should be considered as part of vegetable products)
- Dairy foods and cream, clotted cream (plain), butter, butterfat, fluid buttermilk (plain), dairy-based desserts
- Egg and (frozen, liquid) egg products
- Fresh pasta and noodles
- Fruit and fermented fruit products
- Frozen desserts and ice cream
- Fermented vegetables (carrot, radish, cucumber-based sauces, etc.) and untreated (fresh) vegetable foods
- Fermented non-heat treated processed meat, poultry, etc.
- Infant formulae, formulae for special medical purposes for infants
- Milk and milk products, including dried foods and flavoured fluid milk drinks
- Salt substitutes
- Unripened cheeses, whey protein cheese
- Wines

This partial classification is also obtained on the basis of the General Standard for Food Additives (GSFA) CODEX STAN 192-1995 (Codex Alimentarius Commission 1995) and the Official Methods of Analysis of AOAC International (AOAC 1980, 1990, 1995).

On the other side, currently available analytical techniques for the determination of lactic acid (and other organic acids) in the above-mentioned categories are substantially as follows:

(1) Spectroscopic systems (use of spectrophotometers) for vegetable and canned foods, fruits and fruit products, frozen desserts and ice creams, milk and dairy products, butterfat, beverages and wines
(2) Colorimetric determinations for eggs
(3) Gas chromatography (GC) for eggs (in particular, this technique is useful if other acids such as succinic acids are present)
(4) Amperometric biosensors
(5) Capillary electrophoresis
(6) High-performance liquid chromatography (HPLC) separations
(7) Normal titration and potentiometric titration analyses.

Some difficulties can occur when other organic acids are present: as an example, acetic acid could be found in the same matrix and its determination might be necessary.

In relation to milk, the historical basis of lactic acid determination may be found in the determination of 'milk acidity': the titration with sodium hydroxide (indicator: phenolphthalein) against milk with added rosaniline (standard). This method measures the so-called milk acidity expressed as lactic acid. Normally, milk should not exhibit more than 0.14% if this amount—phosphates, carbon dioxide, casein, citric acid—is not heavily augmented by means of lactose acidification to lactic acid (normal value is 0.002%) (FAO 1986). Lactic acid can be also transformed in acetaldehyde; the resulting compound can show red colours with *p*-hydro-xydiphenyl in strong acid conditions (also good for the determination of dried milk lactate). The intensity of this colour can be used for the spectrophotometric determination. Another AOAC procedure concerns the use of ferric chloride and the development of a coloured solution (FAO 1986). Anyway, spectrophotometry is a good system as reported in certain ambits (Dias et al. 2010).

With reference to flours, acidity can be expressed as lactic acid and usually determined by titration with sodium hydroxide in the aqueous extract of examined flours. Paper chromatography was indicated as a good quantitative test when speaking of different organic acids in fruit juices (FAO 1986).

Anyway, traditional titration is not sufficiently reliable when a little amount of a single acid—such as lactic acid—has to be determined (Casella and Gatta 2002). In addition, acidity may be not always represented by lactic acid alone. At present, available techniques for the determination of lactic acid in foods include also HPLC systems (Andersson and Hedlund 1983). Naturally, these techniques are extremely useful when speaking of different organic acids together (ossalic, malic, lactic, acetic, and other acids) and the need of a concomitant analysis. Food matrices can be of various origins, from fermented vegetables to fermented shrimp waste, and other materials also (Casella and Gatta 2002; Nisperos-Carriedo et al. 1992; Palmer and List 1973; Sánchez-Machado et al. 2008). The key of success of HPLC determination should be found in the coupling of HPLC with ultraviolet detection, refractive index, and pulsed electrochemical detection (Chen et al. 2006; Kordiš-Krapež et al. 2001; Vázquez-Oderiz et al. 1994). Also, HPLC systems are well known for their excellent performances in terms of rapidity, selectivity, sensitivity at least (Sánchez-Machado et al. 2008). Good results have been also reported when speaking of GC (Escobal et al. 1998; Pessione et al. 2015), amperometric biosensors, enzymatic kits (Amer et al. 2015; Pessione et al. 2015), allowing the identification of related lactic acid isomers, and capillary electrophoresis in certain matrices (Buchberger et al. 1997; Goriushkina et al. 2009). Capillary elec-trophoresis has been defined probably a good competitor against HPLC because of claimed lower costs, rapidity, and reduction of waste (de Sena Aquino et al. 2015). However, HPLC and GC methods appear to be one of the most chosen options at present, including also possible separation of lactic acid enantiomers. Anyway, the key of the problem seems to be reliable determination of lactic acid among other present organic acids in the same matrix, while the single determination of this compound without the interference of other acids does not seem to be reported with sufficient frequency in the current literature studies.

Unfortunately, different food matrices often mean different operative conditions. As a consequence, the description of HPLC or GC analyses for lactic acid—among the complex of all possible organic acids—is not possible. The reader is invited to consult more detailed reference literature when speaking of peculiar food matrices.

At present, it may be affirmed that chromatographic techniques and other analytical systems may be performed with good or excellent results in many food matrices, especially when lactic acid is found (or added) in combination with other organic acids.

References

Andersson R, Hedlund B (1983) HPLC analysis of organic acids in lactic acid fermented vegetables. Z Lebensm Unters Forsch 176(6):440–443

Amer MA, Novoa-Díaz D, Puig-Pujol A, Capdevila J, Chávez JA, Turó A, García-Hernández MJ, Salazar J (2015) Ultrasonic velocity of water–ethanol–malic acid–lactic acid mixtures during the malolactic fermentation process. J Food Eng 149:61–69. doi:10.1016/j.jfoodeng.2014.09.042

AOAC (1980) Official methods of analysis of AOAC International, 13th edn. AOAC International, Gaithersburg

AOAC (1990) Official methods of analysis of AOAC International, 15th edn. AOAC International, Gaithersburg

AOAC (1995) Official methods of analysis of AOAC International, 16th edn. AOAC International, Gaithersburg

Buchberger W, Klampfl CHW, Eibensteiner F, Buchgraber K (1997) Determination of fermenting acids in silage by capillary electrophoresis. J Chromatog A 766(1–2):197–203. doi:10.1016/S0021-9673(96)01015-1

Casella IG, Gatta M (2002) Determination of aliphatic organic acids by high-performance liquid chromatography with pulsed electrochemical detection. J Agric Food Chem 50(1):23–28. doi:10.1021/jf010557d

Chen SF, Mowery RA, Castleberry VA, van Walsum GP, Chambliss CK (2006) High-performance liquid chromatography method for simultaneous determination of aliphatic acid, aromatic acid and neutral degradation products in biomass pretreatment hydrolysates. J Chromatogr A 1104(1–2):54–61. doi:10.1016/j.chroma.2005.11.136

Codex Alimentarius Commission (1995) General standard for food additives (GSFA) CODEX STAN 192–1995, last amendment: 2016. Food and Agriculture Organization of the United Nations (FAO), Rome, and World Health Organization (WHO), Geneva

de Sena Aquino AC, Azevedo MS, Ribeiro DH, Costa AC, Amante ER (2015) Validation of HPLC and CE methods for determination of organic acids in sour cassava starch wastewater. Food Chem 172:725–730. doi:10.1016/j.foodchem.2014.09.142

Dias ACB, Silva RAO, Arruda MAZ (2010) A sequential injection system for indirect spectro-photometric determination of lactic acid in yogurt and fermented mash samples. Microchem J 96(1):151–156. doi:10.1016/j.microc.2010.02.016

Escobal A, Iriondo C, Laborra C, Elejalde E, Gonzalez I (1998) Determination of acids and volatile compounds in red Txakoli wine by high performance liquid chromatography and gas chromatography. J Chromatog A 823(1–2):349–354. doi:10.1016/S0021-9673(98)00468-3

FAO (1986) Manuals of food quality control. 8. Food analysis: quality, adulteration and tests of identify. FAO Food and Nutrition papers 14/8, p. 20–21. Swedish International development authority (SIDA), and Food and Agriculture Organization of the United Nations (FAO), Rome

Goriushkina TB, Soldatkin AP, Dzyadevych SV (2009) Application of amperometric biosensors for analysis of ethanol, glucose, and lactate in wine. J Agric Food Chem 57(15):6528–6538. doi:10.1021/jf9009087

Kordiš-Krapež M, Abram V, Kač M, Ferjančić S (2001) Determination of organic acids in white wines by RP-HPLC. Food Technol Biotechnol 39(2):93–99

Nisperos-Carriedo MO, Buslig BS, Shaw PE (1992) Simultaneous detection of de-hydroascorbic, ascorbic and some organic acids in fruits and vegetables by HPLC. J Agric Food Chem 40 (7):1127–1130. doi:10.1021/jf00019a007

Palmer IK, List DM (1973) Determination of organic acids in foods by liquid chromatography. J Agric Food Chem 21(5):903–906

Pessione A, Bianco GL, Mangiapane E, Cirrincione S, Pessione E (2015) Characterization of potentially probiotic lactic acid bacteria isolated from olives: evaluation of short chain fatty acids production and analysis of the extracellular proteome. Food Res Int 67:247–254. doi:10. 1016/j.foodres.2014.11.029

Sánchez-Machado DI, López-Cervantes J, Martínez-Cruz O (2008) Quantification of organic acids in fermented shrimp waste by HPLC. J Food Technol Biotechnol 46(4):456–460

Vazquez Oderiz ML, Vazquez Blanco ME, Lopez Hernandez J, Simal Lozano J, Romero Rodriguez MA (1994) Simultaneous determination of organic acids and vitamin C in green beans by liquid chromatography. J AOAC Int 77(4):1056–1059

Chapter 5
Lactic Acid and Lactic Acid Bacteria: Current Use and Perspectives in the Food and Beverage Industry

Abstract Lactic acid bacteria are widely spread throughout the environment, being symbiotic to humans and among the most important microorganisms used in food fermentations. Despite being a heterogeneous group, lactic acid bacteria share common fermentative pathways, which lead primarily to the production of lactic acid. Their presence in food may be both beneficial and harmful, as their metabolic pathways may also lead to spoilage of certain foods. Furthermore, these microorganisms have gained particular attention due to production of substances of protein structure characterised by an antimicrobial activity (i.e. bacteriocins). These substances are being currently studied for their high potential in the application in food industry for biopreservation, being 'Generally Recognised As Safe'. Therefore, the role of lactic acid bacteria in the food industry is evolving and promising an always increasing number of applications.

Keywords Bacteriocin · Carnobacterium · Lactic acid · Lactobacillus · Lactococcus · Leuconostoc · Streptococcus

Abbreviations

ATP	Adenosine triphosphate
BA	Biogenic amine
CO_2	Carbon dioxide
H_2O_2	Hydrogen peroxide
LAB	Lactic acid bacteria

5.1 Introduction to Lactic Acid Bacteria

Lactic acid bacteria (LAB) are a heterogeneous group, comprising diverse species of Gram-positive chemo-organotrophic bacteria, which share common metabolic and physiological characteristics. Actually, LAB main genera are as follows: *Lactococcus, Lactobacillus, Carnobacterium, Enterococcus, Leuconostoc,*

© The Author(s) 2017
S.M. Ameen and G. Caruso, *Lactic Acid in the Food Industry*,
Chemistry of Foods, DOI 10.1007/978-3-319-58146-0_5

Pediococcus, *Streptococcus*, *Oenococcus*, *Paralactobacillus*, *Tetragenococcus*, *Vagococcus*, and *Weissella*. The LAB group shares the common characteristic of producing lactic acid from hexoses. The most parts of them are strictly fermentative life forms and hence are used in the food industry for various applications. Their fermentative abilities are exploited for instance as starter cultures for industrial processing of fermented dairy, meat, cereal, and vegetable products. They are devoid of cytochromes and derive energy by phosphorylation of substrates, during the oxidation of carbohydrates, not being able to exploit the Krebs Cycle.

5.2   Metabolic Pathways and Fermentation

Fermentation is a metabolic process alternative to respiration, where release of energy occurs through partial oxidation of carbohydrates, with organic compounds produced as the final electron acceptors from the breakdown of the first compounds.

LAB fermentation is considered as a safe preservation technology which enhances food safety, improves organoleptic quality, and helps nutrient absorption, by breaking down proteins, carbohydrates, and fats. Moreover, it extends food shelf life and the preservation of the nutritious components (Widyastuti et al. 2014). The LAB action in fermentations may be either spontaneous or controlled by inoculation of starter cultures.

LAB are mainly subdivided in two categories, based on end products formed during fermentation of glucose: homofermentative (homolactic) life forms and heterofermentative (heterolactic) species. The first group includes bacteria such as *Lactococcus*, *Pediococcus*, and some *Lactobacillus* species (e.g. *L. casei*, *L. delbrueckii*), which obtain a single compound, lactic acid, during fermentation. On the other side, heterofermentative species such as *Leuconostoc*, *Weissella*, and *Carnobacterium* spp. obtain equimolar amounts of lactic acid, ethanol, and carbon dioxide CO_2 (Baglio 2014; Jay 2000). Homolactic species are able to extract a double amount of energy from a given quantity of glucose with respect to heterolactics, even if the homofermentative feature may shift for some strains depending on growth conditions (glucose concentration, pH, nutrient limitation for obligate and facultative homofermenters, etc.) (Brown and Collins 1977). On the other hand, heterolactic life forms are more important in the production of flavour and aroma during fermentation.

The two metabolic pathways differ from biochemical aspects, as homolactic species have the aldolase and hexose isomerase enzymes, while the heterolactics possess phosphoketolase. Therefore, homolactic LAB catabolise one mole of glucose in the Embden–Meyerhof–Parnas pathway, yielding two moles of pyruvate. Intracellular redox balance is retained through the oxidation of nicotinamide adenine dinucleotide, at the same time with pyruvate reduction to lactic acid. This process yields two moles of adenosine triphosphate (ATP) per mole of glucose consumed. On the other side, heterolactic LAB employ the pentose phosphate pathway (or hexose monophosphate shunt) yielding one mole of CO_2 and one mole of ATP, besides lactate and ethanol (Figs. 2.2 and 2.3) (Jay 2000).

Lactobacilli are usually more resistant to acidic pH than other LAB, being able to grow at pH values as low as 3.5–3.8 (Lourens-Hattingh and Viljoen 2001). This capacity represents an ecologically competitive advantage, as they continue to grow during fermentation when the pH has dropped too low for other LAB; hence, they are often in charge for the final stages of many lactic acid fermentations.

Numerous food products owe their existence, features, and palatability to LAB fermentations, such as yoghurt, ripened cheeses, fermented sausages, sauerkraut, and pickles. As for vegetable products, lactic acid fermentations are performed under three basic condition types: dry salted, brined, and non-salted procedures.

Salting, in fact, may help providing a suitable environment for LAB to grow. For instance, Sauerkraut (acid cabbage) is the result of the fermentation of *Leuconostoc mesenteroides*, an acid- and gas-producing coccus. Eventually, *L. plantarum* and *L. brevis* take over when the pH lowers, continuing to produce acid till achieving a pH value of 3, which is too low for lactobacilli also (Plengvidhya et al. 2007). Pickled cucumbers are another fermented product (brine is used instead of dry salt). Some vegetables are fermented by LAB without the prior addition of salt or brine. Examples of non-salted products include *gundruk* (consumed in Nepal), *sinki*, and other wilted fermented leaves. Also, various meat products, such as sausages are processed through fermentations, by means of *Lactobacillus* species. *L. plantarum* is used for instance in starter cultures for the production of pepperoni and Italian salami.

Yeasts are sometimes used in combination with LAB: sourdough bread is made by the fermentation of dough using a wide variety of lactobacilli (*L. sanfranciscensis* and *L. brevis* being the most typical) and yeasts (Corsetti and Settanni 2007).

Nevertheless, the most famous LAB fermented foods are dairy products: cheeses, butter, buttermilk, and yoghurt. Starter cultures include homolactic species (mainly *L. lactis* subsp. *lactis*), capable of converting lactose to lactic acid, and in certain cases heterolactics also (*Leuconostoc mesenteroides* subsp. *cremoris*), depending on the required flavour (Tamime 2002). As for yoghurt, starter cultures are represented by mixed culture of *Streptococcus salivarius* subsp. *thermophilus* and *L. delbueckii* subsp. *bulgaricus*. These two species are responsible together for a much higher production of lactic acid than if they were present singularly: in fact, they cooperate initially enhancing the growth of each other by providing useful metabolites. Then, *L. delbrueckii* begins to dominate as acidity increases, halting *S. salivarius* growth (Courtin and Rul 2003). These two species often are cocultured with other lactobacilli for a probiotic function, as *L. acidophilus* and *L. casei*.

5.3   Properties of LAB

The importance of LAB in food products is not limited to fermentation, as they fulfil many more relevant functions in the food industry. Firstly, they contribute to flavours and textural properties of food products. LAB play an essential role in dairy products, developing for instance individual features of each cheese variety, depending both on the starter and non-starter cultures during the early stages and later on during ripening.

LAB are involved in flavour formation through the metabolism of lactate, fat (i.e. lipolysis), and proteins (i.e. proteolysis). Proteolysis has a central role: besides flavour development, it liberates free amino acids for secondary metabolic pathways and it causes also textural changes in the curd with decrease in water activity through water binding by newly formed carboxyl and amino groups.

Starter cultures depend mainly on their proteolytic metabolism for early growth in milk substrate. In fact, milk contains only few small peptides and free amino acids, and hence, some LAB possess an extracellular, cell-wall-bound proteinase that allows the production of smaller peptides from larger ones originated from casein. These LAB are also able to uptake these hydrolysed peptides intracellularly through oligopeptide and specific amino acid transport systems, di- and tri-peptide transporters (Smit et al. 2005). LAB then possess a wide system of intracellular proteinases/peptidases for further proteolysis (aminopeptidases, endopeptidases, prolinases, di/tri-peptidases, and carboxypeptidases), which can be released in the medium after cell lysis (McSweeney and Sousa 2000). Firstly, most amino acids are converted by aminotransferases to the corresponding α-keto acids, in order to obtain important metabolic intermediates to be converted in hydroxyl acids, aldehydes, and coenzyme A-esters (Van Kranenburg et al. 2002). The oxidative decarboxylation of α-keto acids by the dehydrogenase complex (keto acid dehydrogenase, dihydrolipoyl transacylase, and lipoamide dehydrogenase) contributes to relevant flavour molecules (carboxylic acids, such as isovaleric acid) or to the formation of precursors of other important molecules such as esters, thioesters, cresol, and skatole (Smit et al. 2005).

Citrate metabolism, which is usually performed by non-starter cultures of lactobacilli, may be also accountable for the production of highly flavoured compounds, as acetate, acetoin, 2,3-butanediol, and diacetyl, as well as 2-butanone (Dimos et al. 1996).

Lipids also play a role in aroma in high-fat foods. Because of the negative oxidation-reduction potential (−250 mV) and the weak contribution of LAB to lipolysis, oxidation of cheese lipids is quite limited. However, it has been studied that *Lactococcus* spp., if present in high numbers of extended periods, may contribute quite extensively to the production of free fatty acids, due to the production of lipases/esterases (Chick et al. 1997). In addition, fatty acids can give esters or thioesters which are both important contributors to flavour, by reacting with alcohols and sulphydryl groups (McSweeney and Sousa 2000).

As for yoghurt production, it is important to balance the populations of both the starter culture strains: *S. salivarius* subsp. *thermophilus* and *L. delbueckii* subsp. *bulgaricus* (approximate ratio is 1:1), as an overgrowth of streptococci leads to a less firm and less acid yoghurt. On the opposite hand, the product may result definitely acid if lactobacilli tend to take over, possibly also with a bitter flavour, due to peptides from proteolysis (Schmidt 2008). Instead, these two strains perform better in symbiosis. Initially, *L. delbueckii* subsp. *bulgaricus* fosters growth of the streptococci by releasing valine, leucine, histidine, and methionine thanks to its metabolism. On its side, *S. salivarius* subsp. *thermophilus* supports lactobacilli by forming formic acid. At the initial phase, this mutual help between the two species allows aromatic compounds to be formed faster than with one or the other of these species only (Kroger 1976).

The typical aroma of yoghurt is characterised mainly by acetaldehyde, even if also acetone, acetoin, and diacetyl are suggested as major flavour compounds in addition to acetic, formic, butanoic, and propanoic acids (Routray and Mishra 2011). Acetaldehyde is a metabolic product of carbohydrate metabolism, but it can also be synthesised through threonine aldolase from threonine. This pathway may be responsible for an increase in acetaldehyde in milk fermentation, especially by streptococci. At the same time, *L. acidophilus* alcohol dehydrogenase activity tends to degrade acetaldehyde over time (Schmidt 2008).

Other compounds were reported to be relevant enough in yoghurt flavour composition, such as 2,3-butanedione, 2,3-pentanedione, dimethyl sulphide, and benzaldehyde (Imhof et al. 1994). Furthermore, as for yoghurt, its texture is sustained by LAB by means of the formation of exopolysaccharides (they act as rheological agents), and also by coagulation, as a result of neutralisation of negative charges on milk proteins due to acidity produced by LAB (Yang et al. 2014a).

Besides being implicated as beneficial microorganisms, LAB may also be responsible for defects and spoilage of certain food and beverages. For instance, the formation of calcium lactate white crystals may occur in cheeses on surfaces, resulting from significant quantities of D(−)-lactate during ripening. Normally, L(+)-lactate is transported through the cytoplasmic membrane and converted to D(−)-lactate, through oxidation by L(+)-lactate dehydrogenase to form pyruvate. The last molecule is then reduced to D(−)-lactate by (−)-lactate dehydrogenase. Calcium D (−)-lactate is characterised by a lower solubility than calcium L(+)-lactate; consequently, certain species (e.g. pediococci), which racemise faster than others, may form crystals in cheese (Ledenbach and Marshall 2009). Moreover, heterofermentative LAB can develop off-flavours and gas in ripened cheeses. Catabolism of amino acids by naturally occurring lactobacilli—for instance, *L. lactis* subsp. *lactis* —can produce small amount of gas in cheeses (Martley and Crow 1993). Methyl alcohol and aldehydes produced by certain *L. lactis* strains were associated with off-flavours (Morales et al. 2003).

S. thermophilus and *L. helveticus* were associated with cracks by CO_2 production through decarboxylation of glutamic acid (Zoon and Allersma 1996). Metabolism of tyrosine by certain strains may form a pink-brown discolouration in ripened cheese, depending on the presence of oxygen on surfaces (Shannon et al. 1977).

Finally, cheese represents an ideal substrate for production of biogenic amines (BA), reaching concentrations up to 2000 mg kg^{-1}, as it is not a sterile food; also, proteolysis of casein guarantees the availability of free amino acids. In fact, bacteria may produce BA due to amino acid decarboxylases which remove the α-carboxyl group from a certain amino acid to give its equivalent amine. The most relevant BA in dairy products are histamine, tyramine (produced by enzymatic decarboxylation of histidine and tyrosine, respectively), and putrescine (Linares et al. 2012).

Consumption of food containing high BA amounts may elicit adverse reactions in susceptible humans, with serious health problems. Non-starter LAB capable of BA formation include *L. bulgaricus*, *L. casei*, *L. acidophilus*, but also some *Lactococcus*, *Leuconostoc*, and *Streptococcus*, strains which are all normal microbiota of dairy products and hence can be potentially present during processing. Interestingly, certain strains of starter microbiota, such as *L. lactis,* have been found to produce BA as well, but screening measures for decarboxylase activity can easily avoid such issue (O'Sullivan et al. 2013).

LAB represent also the most famous group among probiotics, a definition referring to viable microorganisms that promote or support a beneficial balance of the autochthonous microbial population of the gastrointestinal tract (Holzapfel et al. 2001). In particular, *L. acidophilus*, *L. casei* and *L. paracasei*, and *Bifidobacterium* spp. have a major role (Holzapfel et al. 1997). Many recent studies have focused on the numerous beneficial effects of probiotics, attaining an increasing interest in consumers. Benefits include not only an improved nutritional composition of the food, by releasing amino acids or vitamins as a result of their metabolism, but also an enhanced host immune system, the inhibition of pathogens, and the reduction of the prevalence of allergy and of symptoms of lactose intolerance. Moreover, probiotics can decrease the risk of certain cancers and control inflammatory bowel diseases (Parvez et al. 2006).

Milk and dairy products are theoretically good substrates for pathogens and other spoilage bacteria, but LAB being able to acidify the medium and protect these products from unwanted bacteria. In fact, lactic acid fermentation inhibits the growth of other non-desirable organisms. Many organic acids are produced as end products of LAB metabolism: for instance, lactic and acetic acids are the main obtained products. They act firstly diffusing through the membranes of target organisms, hence reducing cytoplasmic pH, so as stopping metabolic activities (Piard and Desmazeaud 1991). Other organic acids, such as propionic acid, were shown to neutralise the plasmic membrane potential and so increase its permeability (Kabara and Eklund 1991).

Some *L. lactis* strains were reported to control mycotoxigenic mould growth (Florianowicz 2001). *L. plantarum* 21B was shown to have antifungal activity because of phenyllactic and 4-hydroxy-phenyllactic acids (Lavermicocca et al. 2003). Also, fatty acids may have a role in acting as potent antifungal compounds such as caproic acid (Corsetti et al. 1998). Hydrogen peroxide (H_2O_2) is also produced by LAB by means of the oxidation of lipid membranes, due to lack of catalase. In some foods (example: milk), H_2O_2 may also react with thiocyanate, forming hypothiocyanite and other intermediate metabolites which hinder the growth of a wide range of unwanted microorganisms (Schnürer and Magnusson 2005).

LAB interact also with the accumulation of certain mycotoxins, as for instance ochratoxin A, fumonisin B_1 and B_2, aflatoxin B_1, and zearalenone (El-Nezami et al. 2002; Fuchs et al. 2008; Gratz et al. 2004; Niderkorn et al. 2006). In fact, toxins are reported to be bound by bacteria, which remove efficiently mycotoxins from the media. *L. casei* was found to be among the strongest aflatoxin binder (Fazeli et al.

2009). This binding interaction could be used as a potential future application, in order to reduce mycotoxin availability in food, also because non-viable LAB maintain their binding capacity (Dalié et al. 2010).

Finally, LAB also decrease the proliferation of unwanted bacteria by the release of bacteriocins, small peptides with antimicrobial properties, which are gaining attention worldwide for their potential applications in various fields.

5.4 Bacteriocins

Bacteriocins, ribosomally synthesised, are antimicrobial proteins with a narrow (against related species) or broad (against non-related species) killing spectrum. Maybe, they could have evolved as a result of space and resource competition with other microbes in the environment, but now, they are extensively being studied primarily for their potentials in food safety, as 'Generally Recognised As Safe' preservatives. In fact, consumers demand for fresh, ready-to-eat, and safe food is increasing worldwide, as well as the demand for minimally preserved food. The toxic effect of many chemical additives has become more popular, and it elicits more consciousness of natural healthy food. LAB bacteriocins are very interesting as they are quite resistant to high thermal stress and to a wide pH range. Moreover, they can be degraded by proteolytic enzymes as a result of their proteinaceous nature, minimising their life both in the environment and in human gastrointestinal tract and hence the probability of the developing antibiotic resistance by target bacteria. Instead, they interact with anionic lipids, located on the membrane of Gram-positive bacteria, and they are able to form pores in the membrane, hence disrupting the cell organisation.

In recent years, new bacteriocins are continuously being discovered, showing that they are a heterogeneous group of diverse compounds, many of them with unique characteristics.

Due to the wide variety of molecules, bacteriocins were divided into three classes, according to their structure and characteristics (Yang et al. 2014b). Post-translationally modified small peptides (<5 kDa) are Class I bacteriocins (lantibiotics). Moreover, they are linear, quite elongated, heat stable, positively charged, and amphipathic molecules, characterised by polycyclic thioether amino acids lanthionine or methyllanthionine, resulting from post-translational modifications, and the unsaturated amino acids dehydroalanine and 2-aminoisobutyric acid. The two major representatives of this class are nisin—the most widely exploited in food industry, and already approved and licensed for utilisation as a preservative by many countries—and lacticin 3147.

Nisin has a wide antimicrobial spectrum against Gram-positive bacteria, including *L. monocytogenes*, in meat products (Pawar et al. 2000) as well as in dairy products (Bhatti et al. 2004). Nisin has also been applied to cheeses in order to control the growth of spores by *Clostridium tyrobutyricum* (Rilla et al. 2003). This bacteriocin is formed by a pentacyclic structure with four β-methyllanthionine and one lanthionine residues. It is heat stable at 121 °C and is resistant to trypsin, pepsin, and carboxyl peptidase, while it is sensitive to α-chymotrypsin (Zacharof and Lovitt 2012).

Lacticin 3147 has also a broad spectrum of activity, inhibiting growth of *Bacillus* sp., *C. tyrobutyricum*, and most mastitis-causing streptococci. Pathogenic bacteria, such as *L. monocytogenes* and *S. aureus*, are sensitive to lacticin 3147 too (Ryan et al. 1996).

Class II is composed of small amino acids (molecular weights are <10 kDa, usually between 30–60 kDa), relatively heat stable, positively charged, and without lanthionine. This class was subdivided in subclasses (Cotter et al. 2013). Class IIa molecules are pediocin-like or listeria-active bacteriocins, which are characterised by the presence of a consensus sequence (Tyr-Gly-Asn-Gly-Val-Xaa-Cys) in the *N*-terminus. They show high degree of homology in their amino acid sequences (40–60%), and the major representatives of this subclass are pediocin PA-1 from strains of *Pediococcus acidilactici* and carnobacteriocin X.

Class IIb is made of two-component bacteriocins as lactacin F, which are two different peptides performing their antimicrobial activity synergistically. Class IIc is composed of circular peptides, as carnocyclin A by *Carnobacterium maltaromaticum*, while Class IId bacteriocins include linear, single-peptide, non-pediocin-like molecules such as lactococcin A.

Class III is characterised by heat labile, large molecular weight peptides (>30 kDa) which are not so well investigated. They are divided in two groups: group A (i.e. bacteriolytic enzymes), including Enterolisin A by *Enterococcus faecalis*, and group B containing non-lytic peptides, as caseicin 80 and helveticin J by *L. casei* and *L. helveticus*, respectively.

Various external factors influence the production of bacteriocins by LAB, as for instance nutrient availability. Lack of nutrients, such as manganese, and the presence of sodium chloride may be limiting factors (Verluyten et al. 2004a, b). Moreover, interaction with phospholipid emulsifiers and other food components (Aasen et al. 2003) or inactivation by formation of nisin-glutathione adduct (Ross et al. 2003) may affect the efficacy of bacteriocins.

Bacteriocin activity is also linked to the physiological state of microorganisms. In fact, actively growing microorganisms are more sensitive to bacteriocins, while endospores that have a reduced metabolic activity are more resistant (Galvez et al. 2007). Activity against Gram-negative bacteria occurs mostly only when the outer membrane is compromised, as for instance after hurdle technology treatments (Stevens et al. 1991). It has been shown that the combination of bacteriocins with other treatments and/or chemical substances (as for instance organic acids and their salts) potentiates the antimicrobial activity. Use of non-thermal treatments such as 'pulsed electric field' technologies, in combination with bacteriocins, has shown very useful in controlling pathogenic bacteria (Calderon-Miranda et al. 1999; Sobrino-Lopez and Martin Belloso 2006), and besides, it does not affect the nutritional quality of food products. Also, modified atmosphere packaging systems act synergistically with bacteriocins in the inactivation of microorganisms, by enhancing membrane permeabilisation (Nilsson et al. 2000).

Recently, another application of bacteriocins has emerged: bioactive packaging helps protecting food from unwanted bacteria. Bioactive packaging can be used with bacteriocin-containing sachets that release the active molecules gradually during food storage, or by direct immobilisation of bacteriocins on the packaging surface. This latter can be achieved through incorporation of bacteriocins into polymers, as biodegradable protein films, through heat press or adsorption of bacteriocin on polyethylene, polypropylene, polyester acrylics, and other polymers (Zacharof and Lovitt 2012).

Bacteriocins from *L. curvatus* immobilised in polyethylene film decreased counts of *L. monocytogenes* during shelf life storage of various types of meat (Ercolini et al. 2006; Mauriello et al. 2004). Alternatively, bacteriocins may be produced in vivo by starter cultures directly in the food.

The potential importance of bacteriocins was identified also for use as antimicrobials in the clinical field. In fact, they are reported to hinder important pathogens in the treatment of some veterinary microbial diseases. Applications range from topical use in the treatment of skin infections to inflammation and ulcers, and to antimicrobial activity in medical devices as catheters (Bower et al. 2002; Goldstein et al. 1998; Sears et al. 1992). Commercially available products are utilised for treatment of mastitis in dairy cattle (Ryan et al. 1998). Furthermore, nisin has already undergone human clinical trials for treatment of *Helicobacter pylori*-associated peptic ulcers (Hancock 1997).

Some bacteriocins such as nisin A and lacticin Q have been reported to have a stronger bactericidal activity on methicillin-resistant *Staphylococcus aureus* planktonic and biofilm cells with respect to the potent antibiotic vancomycin (Okuda et al. 2013). It is then of the uttermost importance to continue studying and exploring the extensive field of bacteriocins, in order to widen our reserve of antimicrobial compounds for a better control of food and clinical pathogens.

References

Aasen IM, Markussen S, Møretrø T, Katla T, Axelsson L, Naterstad K (2003) Interactions of the bacteriocins sakacin P and nisin with food constituents. Int J Food Microbiol 87(1–2):35–43. doi:10.1016/S0168-1605(03)00047-3

Baglio E (2014) Chemistry and technology of yoghurt fermentation. Springer, SpringerBriefs in Chemistry of Foods

Bhatti M, Veeramachaneni A, Shelef LA (2004) Factors affecting the antilisterial effects of nisin in milk. Int J Food Microbiol 97(2):215–219. doi:10.1016/j.ijfoodmicro.2004.06.010

Bower CK, Parker JE, Higgins AZ, Oest ME, Wilson JT, Valentine BA, Bothwell MK, McGuire J (2002) Protein antimicrobial barriers to bacterial adhesion: in vitro and in vivo evaluation of nisin-treated implantable materials. Coll Surf B Biointerfer 25(1):81–90. doi:10.1016/S0927-7765(01)00318-6

Brown WV, Collins EB (1977) End products and fermentation balances for lactic streptococci grown aerobically on low concentration of glucose. Appl Environ Microbiol 33(1):38–42

Calderón-Miranda ML, Barbosa-Canovas GV, Swanson BG (1999) Inactivation of *Listeria innocua* in liquid whole egg by pulsed electric fields and nisin. Int J Food Microbiol 51(1):7–17. doi:10.1016/S0168-1605(99)00070-7

Chick JF, Marchesseau K, Gripon JC (1997) Intracellular esterase from *Lactococcus lactis* subsp. *lactis* NCDO 763: purification and characterization. Int Dairy J 7(2–3):169–174. doi:10. 1016/S0958-6946(97)00001-0

Corsetti A, Settanni L (2007) Lactobacilli in sourdough fermentation. Food Res Int 40(5):539–558. doi:10.1016/j.foodres.2006.11.001

Corsetti A, Gobetti M, Rossi J, Damiani P (1998) Antimould activity of sourdough lactic acid bacteria: identification of a mixture of organic acids produced by *Lactobacillus sanfrancisco* CB1. Appl Microbiol Biot 50(2):253–256

Cotter PD, Ross RP, Hill C (2013) Bacteriocins— a viable alternative to antibiotics? Nat Rev Microbiol 11(2):95–105. doi:10.1038/nrmicro2937

Courtin P, Rul FO (2003) Interactions between microorganisms in a simple ecosystem: yogurt bacteria as a study model. Le Lait 84(1–2):125–134. doi:10.1051/lait:2003031

Dalié DKD, Deschamps AM, Richard-Forget F (2010) Lactic acid bacteria—potential for control of mould growth and mycotoxins: a review. Food Control 21(4):370–380. doi:10.1016/j. foodcont.2009.07.011

Dimos A, Urbach GE, Miller AJ (1996) Changes in flavour and volatiles of full-fat and low-fat cheeses during maturation. Int Dairy J 6:981–995

El-Nezami HS, Polychronaki N, Salminen S, Mykkänen H (2002) Binding rather metabolism may explain the interaction of two food-grade *Lactobacillus* strains with zearalenone and its derivative a-zearalenol. Appl Environ Microbiol 68(7):3545–3549. doi:10.1128/AEM.68.7. 3545-3549.2002

Ercolini D, Storia A, Villani F, Mauriello G (2006) Effect of a bacteriocins activated polythene film on *Listeria monocytogenes* as evaluated by viable staining and epifluorescence microscopy. J Appl Microbiol 100(4):765–772. doi:10.1111/j.1365-2672.2006.02825.x

Fazeli MR, Hajimohammadali M, Moshkani A, Samadi N, Jamalifar H, Khoshayand MR, Vaghari E, Pouragahi S (2009) Aflatoxin B1 binding capacity of autochthonous strains of lactic acid bacteria. J Food Protection 72(1):189–192

Florianowicz T (2001) Antifungal activity of some microorganisms against *Penicillium expansum*. Eur Food Res Technol 212(3):282–286. doi:10.1007/s002170000261

Fuchs S, Sontag G, Stidl R, Ehrlich V, Kundi M, Knasmüller S (2008) Detoxification of patulin and ochratoxin A, two abundant mycotoxins, by lactic acid bacteria. Food Chem Toxicol 46 (4):1398–1407. doi:10.1016/j.fct.2007.10.008

Gálvez A, Abriouel H, López RL, Ben Omar N (2007) Bacteriocin-based strategies for food biopreservation. Int J Food Microbiol 120(1–2):51–70. doi:10.1016/j.ijfoodmicro.2007.06.001

Goldstein BP, Wei J, Greenberg K, Novick R (1998) Activity of nisin against *Streptococcus pneumoniae*, in vitro, and in mouse infection model. J Antimic Chemother 42(2):277–278

Gratz S, Mykkänen H, Ouwehand AC, Juvonen R, Salminen S, El-Nezami H (2004) Intestinal mucus alters the ability of probiotic bacteria to bind aflatoxin B1 in vitro. Appl Environ Microbiol 70(10):6306–6308. doi:10.1128/AEM.70.10.6306-6308.2004

Hancock REW (1997) Peptide antibiotics. Lancet 349(9049):418–420. doi:10.1016/S0140-6736 (97)80051-7

Holzapfel WH, Schillinger U, Du Toit M, Dicks L (1997) Systematics of probiotic lactic acid bacteria. Microecol Ther 26:1–10

Holzapfel WH, Haberer P, Geisen R, Björkroth J, Schillinger U (2001) Taxonomy and important features of probiotic microorganisms in food and nutrition. Am J Clin Nutr 73(2 Suppl):365S–373S

Imhof R, Glatti H, Bosset JO (1994) Volatile organic compounds produced by thermophilic and mesophilic single strain dairy starter cultures. Lebensm-Wiss-Technol 28(1):78–86. doi:10. 1016/S0023-6438(95)80016-6

Jay JM (2000) Fermentation and fermented dairy products. In: Jay JM (ed) Modern food microbiology, 6th edn. Aspen Publication, Springer US, New York, pp. 113–130. doi:10.1007/ 978-1-4615-4427-2_7

Kabara J, Eklund T (1991) Organic acid and esters. In: Russel NJ, Gould GW (eds), Food Preservatives. Blackie, Glasgow and London, pp. 44–71

Kroger M (1976) Quality of yogurt. J Dairy Sci 59(2):344–50

Lavermicocca P, Valerio F, Visconti A (2003) Antifungal activity of phenyllactic acid against molds isolated from bakery products. Appl Environ Microbiol 69(1):634–640. doi:10.1128/AEM.69.1.634-640.2003

Ledenbach LH, Marshall RT (2009) Microbiological spoilage of dairy products. In: Sperber WH, Doyle MP (eds) Compendium of the microbiological spoilage of food and beverages. Springer, New York, pp 41–67

Linares DM, Del Río B, Ladero V, Martínez N, Fernández M, Martín MC, Alvarez MA (2012) Factors influencing biogenic amines accumulation in dairy products. Front Microbiol 3:180. doi:10.3389/fmicb.2012.00180

Lourens-Hattingh A, Viljoen BC (2001) Yogurt as probiotic carrier food. Int Dairy J 11(1–2):1–17. doi:10.1016/S0958-6946(01)00036-X

Martley FG, Crow VL (1993) Interactions between non-starter microorganisms during cheese manufacture and ripening. Int Dairy J 3(4):641–464. doi:10.1016/0958-6946(93)90027-W

Mauriello G, Ercolini D, La Storia A, Casaburi A, Villani F (2004) Development of polythene films for food packaging activated with an antilisterial bacteriocin from *Lactobacillus curvatus* 32Y. J Appl Microbiol 97(2):314–322. doi:10.1111/j.1365-2672.2004.02299.x

McSweeney PLH, Sousa MJ (2000) Biochemical pathways for the production of flavour compounds in cheeses during ripening: a review. Lait 80(3):293–324. doi:10.1051/lait:2000127

Morales P, Fernandez-Garcia E, Gaya P, Nunez M (2003) Formation of volatile compounds by wild *Lactococcus lactis* strains isolated from raw ewes' milk cheese. Int Dairy J 13(2–3):201–209. doi:10.1016/S0958-6946(02)00151-6

Niderkorn V, Boudra H, Morgavi DP (2006) Binding of *Fusarium* mycotoxins by fermentative bacteria in vitro. Appl Environ Microbiol 101(4):849–856. doi:0.1111/j.1365-2672.2006.02958.x

Nilsson L, Chen Y, Chikindas ML, Huss HH, Gram L, Montville TJ (2000) Carbon dioxide and nisin act synergistically on *Listeria monocytogenes*. Appl Environ Microbiol 66(2):769–774

Okuda K, Zendo T, Sugimoto S, Iwase T, Tajima A, Yamada S, Sonomoto K, Mizunoe Y (2013) Effects of bacteriocins on methicillin-resistant *Staphylococcus aureus* biofilm. Antimicrob Agents Chemother 57(11):5572–5579. doi:10.1128/AAC.00888-13

O'Sullivan DJ, Giblin L, McSweeney PL, Sheehan JJ, Cotter PD (2013) Nucleic acid-based approaches to investigate microbial-related cheese quality defects. Front Microbiol 4:1. doi:10.3389/fmicb.2013.00001

Parvez S, Malik KA, Ah Kang S, Kim HY (2006) Probiotics and their fermented food products are beneficial for health. J Appl Microbiol 100(6):1171–1185. doi:10.1111/j.1365-2672.2006.02963.x

Pawar DD, Malik SVS, Bhilegaonkar KN, Barbuddhe SB (2000) Effect of nisin and its combination with sodium chloride on the survival of *Listeria monocytogenes* added to raw buffalo meat mince. Meat Sci 56(3):215–219. doi:10.1016/S0309-1740(00)00043-7

Piard JC, Desmazeaud M (1991) Inhibition factors produced by lactic acid bacteria: Oxygen metabolites and catabolism end-products. Lait 71(5):525–541. doi:10.1051/lait:1991541

Plengvidhya V, Breidt F, Lu Z, Fleming HP (2007) DNA fingerprinting of lactic acid bacteria in sauerkraut fermentations. Appl Env Microbiol 73(23):7697–7702. doi:10.1128/AEM.01342-07

Rilla N, Martínes-Delgado T, Rodríguez A (2003) Inhibition of *Clostridium tyrobutyricum* in Vidiago cheese by *Lactococus lactis* ssp. *lactis* IPLA 729, a nisin Z producer. Int J Food Microbiol 85(1–2):23–33. doi:10.1016/S0168-1605(02)00478-6

Ross AIV, Griffiths MW, Mittal GS, Deeth HC (2003) Combining nonthermal technologies to control foodborne microorganisms. Int J Food Microbiol 89(2–3):125–138. doi:10.1016/S0168-1605(03)00161-2

Routray W, Mishra HN (2011) Scientific and technical aspects of yogurt aroma and taste: a review. Compr Rev Food Sci Food Saf 10(4):208–220. doi:10.1111/j.1541-4337.2011.00151.x

Ryan MP, Rea MC, Hill C, Ross RP (1996) An application in cheddar cheese manufacture for a strain of *Lactococcus lactis* producing a novel broad-spectrum bacteriocin, lacticin 3147. Appl Environ Microbiol 62(2):612–619

Ryan MP, Meaney WJ, Ross RP, Hill C (1998) Evaluation of lacticin 3147 and a teat seal containing this bacteriocin for inhibition of mastitis pathogens. Appl Environ Microbiol 64 (6):2287–2290

Schmidt RH (2008) Microbiological considerations related to dairy processing. In: Chandan RC (ed) Dairy processing and quality assurance, pp. 105–143. Wiley-Blackwell, Oxford. doi:10.1002/9780813804033.ch5

Schnürer J, Magnusson J (2005) Antifungal lactic acid bacteria as biopreservatives. Trends Food Sci Tech 16(1–3):70–78. doi:10.1016/j.tifs.2004.02.014

Sears PM, Smith BS, Stewart WK, Gonazalez RN (1992) Evaluation of a nisin-based germicidal formulation on teat skin of live cows. J Dairy Sci 75(11):3185–3190. doi:10.3168/jds.S0022-0302(92)78083-7

Shannon EL, Olson NF, Deibel RH (1977) Oxidative metabolism of lactic acid bacteria associated with pink discoloration in Italian cheese. J Dairy Sci 60(11):693–1697. doi:10.3168/jds.S0022-0302(77)84092-7

Smit G, Smit BA, Engels WJ (2005) Flavour formation by lactic acid bacteria and biochemical flavour profiling of cheese products. FEMS Microbiol Rev 29(3):591–610

Sobrino-Lopez A, Martin Belloso O (2006) Enhancing inactivation of *Staphylococcus aureus* in skim milk by combining high-intensity pulsed electric fields and nisin. J Food Prot 69(2):345–353

Stevens KA, Sheldon BW, Klapes NA, Klaenhammer TR (1991) Nisin treatment for inactivation of *Salmonella* species and other gram negative bacteria. Appl Environ Microbiol 57(12):3613–3615

Tamime AY (2002) Microbiology of starter cultures. In: Robinson RK (ed) Dairy microbiology handbook, 3rd edn. John Wiley & Sons, New York, pp 261–367

Van Kranenburg R, Kleerebezem M, van Hylckama Vlieg JET, Ursing BM, Boekhorst J, Smit BA, Ayada EHE, Smita G, Siezen RJ (2002) Flavour formation from amino acids by lactic acid bacteria: predictions from genome sequence analysis. Int Dairy J 12(2–3):111–121. doi:10.1016/S0958-6946(01)00132-7

Verluyten J, Messens W, De Vuyst L (2004a) Sodium chloride reduces production of curvacin A, a bacteriocin produced by *Lactobacillus curvatus* strain LTH 1174, originating from fermented sausage. Appl Environ Microbiol 70(4):2271–2278. doi:10.1128/AEM.70.4.2271-2278.2004

Verluyten J, Leroy F, De Vuyst L (2004b) Influence of complex nutrient source on growth of and curvacin a production by sausage isolate *Lactobacillus curvatus* LTH 1174. Appl Environ Microbiol 70(9):5081–5088. doi:10.1128/AEM.70.9.5081-5088.2004

Widyastuti Y, Rohmatussolihat Febrisiantosa A (2014) The role of lactic acid bacteria in milk fermentation. Food Nutr Sci 5(4):435–442. doi:10.4236/fns.2014.54051

Yang TX, Wu KY, Wang F, Liang XL, Liu QS, Li G, Li QY (2014a) Effect of exopolysaccharides from lactic acid bacteria on the texture and microstructure of buffalo yoghurt. Int Dairy J 34 (2):252–256. doi:10.1016/j.idairyj.2013.08.007

Yang SC, Lin CH, Sung CT, Fang JY (2014b) Antibacterial activities of bacteriocins: application in foods and pharmaceuticals. Front Microbiol 5:241. doi:10.3389/fmicb.2014.00241

Zacharof MP, Lovitt RW (2012) Bacteriocins produced by lactic acid bacteria: a review article. APCBEE Proc 2:50–56. doi:10.1016/j.apcbee.2012.06.010

Zoon P, Allersma D (1996) Eye and crack formation in cheese by carbon dioxide from decarboxylation of glutamic acid. Netherlands Milk Dairy J 50(2):309–318